Transmission Lines with Pulse Excitation

ELECTRICAL SCIENCE
A Series of Monographs and Texts

Editors: Henry G. Booker and Nicholas DeClaris
A complete list of titles in this series appears at the end of this volume

TRANSMISSION LINES WITH PULSE EXCITATION

Georges Metzger

École Centrale des Arts
et Manufactures de Paris
Compagnie Bull-General Electric
Paris, France

Jean-Paul Vabre

École Superieure d'Electricité
Conservatoire National des Arts et Métiers
Compagnie Bull-General Electric
Paris, France

Translated by Robert McDonough

Department of Electrical Engineering
University of Delaware
Newark, Delaware

1969

ACADEMIC PRESS New York and London

First published in the French language under the title "Electronique des Impulsions, Tome II. Circuits à Constantes Réparties," in the series Les Techniques de Base de l'Informatique published under the auspices of the Compagnie Bull-General Electric by Masson et Cie, Paris, © 1966.
("Electronique des Impulsions, Tome I. Circuits à Constantes Localisées," was published by Masson and Cie., Paris, © 1966.)

ENGLISH LANGUAGE EDITION

ACADEMIC PRESS, INC.
111 Fifth Avenue, New York, New York 10003

United Kingdom Edition published by
ACADEMIC PRESS, INC. (LONDON) LTD.
Berkeley Square House, London W1X 6BA

LIBRARY OF CONGRESS CATALOG CARD NUMBER: 69-18342

PRINTED IN THE UNITED STATES OF AMERICA

Foreword

When the various devices of modern technology are analyzed from a very general point of view, whether it be a large particle accelerator, a radar system, a guided missile, or a digital computer, it is found to be very fruitful to consider these complex devices as an assemblage of a certain number of simple units, each transforming "signals" according to a particular "function." The term *signal* is used very broadly here. For example, we would say that the temperature of an enclosure is the input signal to a thermometer, and that the height of the mercury column is the output signal.

An engineer who wishes to design or analyze a complex system thus must have at his disposal an appropriate formalism for describing concisely the signals which he encounters, and the *transformations* carried out on these signals by the various elements of the system. This formalism takes the form of mathematical tools adapted to one or another class of systems.

The devices of the nineteenth century, and of the beginning of the twentieth, especially in electrical engineering and radio, made great use of *sinusoidal* signals, for which a large mathematical apparatus had been developed during the nineteenth century (essentially the use of imaginary numbers and of exponential functions with imaginary argument, followed by use of the Fourier transform).

Pulse signals had been encountered earlier, notably in telegraphy, but it was only around 1940 that the study of radar forced the use of a mathematical apparatus adapted to this type signal, and led to its popular use. This apparatus is the *Laplace transform*, which rests upon the notion of an oscillation with complex frequency.

In the natural course of its development, this new formalism invaded the young theory of servomechanisms, encountering great success and enriching itself with new practical methods, and a little later was generalized to all areas of science and engineering in which pulse phenomena had been studied, and in particular to the technology of electronic calculators.

The very rapid growth in the use of pulse electronics explains the discrepancy which has come about in the past twenty years between the needs of engineers on the one hand, and the small amount of literature and small number of courses available on this subject in many countries on the other.

The present work treats some problems of transmission lines which have become of great importance, because of the extremely short switching times of certain component devices in the pulse regime. The round-trip time of a pulse through a cable 10 cm long is on the order of one nanosecond; but one nanosecond is not a negligible time with respect to the switching time of a diode or a fast transistor, or of certain memory elements (thin magnetic films, superconducting films). Thus the perturbations caused by very short interconnections can no longer be neglected, if it is desired that these interconnections not be the limiting factor in the speed of operation of electronic devices.

To be sure, to avoid these difficulties, "integrated circuits" have been developed, the extremely small size of which solves the problem of internal connections. But there will always need to be external connections between modules, or connections between modules and peripheral devices, and it is desirable to study carefully the perturbations which these connections can cause.

We will feel that this book will have transmitted its message, if in addition to contributing to instruction in engineering, it aids in the dialogue between the universities and industry which is needed in the development of engineers.

P. Davous

Engineer, École Polytechnique and Génie Maritime
Research Director, Compagnie Bull-General Electric

Preface

The behavior of a signal on a transmission line is described by a linear partial differential equation, the celebrated *telegrapher's equation*. The derivation of this equation should be only the introduction to a study of transmission lines; the important thing is its solution. The radio engineer, confronted with this problem, has attacked it pragmatically. Rather than search for a very general solution, probably of little use, he has sought as complete a solution as is possible for the signals used in practice.

The sinusoid is an example of a "useful" signal. The existence of sinusoidal oscillators covering a large range of frequencies, and the development of modulators allowing information to be placed on such a signal, have led to the complete development of the solution of the telegrapher's equation for a sinusoidal time function. Numerous studies are devoted to this aspect of the question. The engineer deals with sinusoids with confidence; he has in reserve a formidable mathematical arsenal consisting of the theory of distributions and the theory of the Fourier integral. He assumes, more or less consciously, that knowledge of the behavior of a circuit under sinusoidal excitation, because of known mathematical results, will allow him to treat more complex cases, in particular, that of pulse excitation. In fact, this impression of security is false. When one actually attempts to deal with problems of this type, as L. Brillouin has, one collides with almost insurmountable difficulties. Complete calculation of the propagation of a pulse on a real line, having both loss and dispersion, is only possible for certain special cases, and then only after long computation. The physical interpretation of these results is also delicate.

Another type of signal is becoming important. This is the pulse, isolated or repeated with a pattern such that the pulse train can carry information. An engineer examining such a signal on the screen of an oscilloscope finds it very difficult to accept the fact that this simple rectangular function can be studied only with the aid of the Fourier integral, and then only after numerous chapters devoted to the propagation of signals which appear to be more complex. Such an engineer will thus try to solve the problem directly, without relying on studies of other signals. This natural reaction has led Jean-Paul Vabre and Georges Metzger, engineers with Compagnie Bull-General Electric, to write this book which I have the pleasure to present.

It is natural to consider first the simplest pulse, the unit step. The linearity of the equations of interest allows any arbitrary pulse to be described as a superposition of unit steps, in effect. If the basic pulse propagates without distortion, the calculations will be simple. For this, it is necessary that the line have no dispersion and no loss, or at least small loss. Fortunately, this is the case with most lines used in pulse circuits. The methods of calculation presented in Chapters II and III are thus quite useful, and allow the behavior of a line under pulse excitation to be described precisely. The agreement between calculated results and actual oscilloscope photographs is excellent, and justifies interest in the methods.

But the study is not limited to lossless transmission lines without dispersion. It has been generalized using the Laplace transform. The propagation of an arbitrary pulse on an arbitrary line can thus be treated; Chapter IV is devoted to the study of this problem. The solution appears in the form of an integral as complicated as the Fourier integral, and in effect, the two formalisms are equivalent in practice. The fact that theoreticians prefer the Fourier integral, which has a rigorous formalism, while engineers more readily use the Laplace transform, should not obscure the equivalence of these methods for the treatment of the problems usually encountered in practice. This is illustrated in the last chapter of the book, which describes the behavior of a line in the sinusoidal regime as a natural consequence of the preceding chapters. Starting from the pulse regime, one is led to and through the sinusoidal regime with the guidance of the Laplace integral, the reverse of the traditional path.

These considerations should not, however, cause one to lose sight of the fact that lines used in practice for the transmission of pulses are chosen such that the signals are not very much deformed. The methods of calculation in this book are thus highly useful, and the reader will be prepared to apply them, especially if he solves the numerous exercises which have been placed at the end of each chapter. In these, the authors have shown that they can unite engineering skills with pedagogical qualities. University students and engineers seeking to bring their knowledge up to date will find in this book ample working tools. But this book should not be regarded solely as a manual for those engaged in practice. The permanent evolution of the techniques poses delicate problems to those engaged in engineering teaching. The reader who has an eye for pedagogy will surely be led to some new techniques, possibly to the betterment of engineering teaching.

M.-Y. BERNARD

Department of Radio Electricity
Conservatoire National des Arts et Métiers

Acknowledgments

We especially thank J. Hug, Compagnie Bull-General Electric, who kindly aided in the completion of this book. We express equally our gratitude to P. Chapouille and J. Henry, for the suggestions which they gave us while we were writing this book.

G. M. and J.-P. V.

Contents

FOREWORD v

PREFACE vii

ACKNOWLEDGMENTS ix

NOTATION xiii

Chapter I. **General Equations for Transmission Lines**

1. Introduction 1
2. Definition of the State of the Line at a Point 3
3. Per Unit Length Parameters, or Primary Parameters 4
4. General Equations 5
5. Use of Operational Calculus (The Laplace Transform) 7
6. Appendix. The Per Unit Length Parameters L, C, R, G 12

Chapter II. **The Method of Traveling Waves**

1. The Method of Traveling Waves for the Lossless Line 21
2. Principal Results 25
3. Examples 29
4. A Tabular Method 51
5. Appendix. Formulas for the Method of Traveling Waves in the General Case of a Line with Losses 57

   Exercises 61

Chapter III. **The Bergeron Diagram**

1. Basic Relations 65
2. Principle of the Method of Bergeron 68
3. The Introduction of Lumped Elements 84
4. Input Signal $e(t)$ Different from a Step Function 92

   Exercises 94

Chapter IV. **Line Transients Using Operational Calculus**

1. Generalities 97
2. Lossless Lines. General Properties 100
3. Lines with Loss in the Beginning of the Transient Regime 120
4. Lossy Lines Satisfying the Heaviside Condition 124
5. The Influence of Skin Effect 125
6. An Equivalent Lumped Circuit for a Line 131
7. Appendix 138
Exercises 146

Chapter V. **Lines in the Sinusoidal Steady State**

1. Preliminary Definitions 149
2. Solution of the Line Equations, and the Secondary Parameters Z_c and γ 151
3. Characteristic Impedance for Some Special Cases 153
4. Interpretation of γ 154
5. Decomposition of the Steady State into Two Traveling Waves 154
6. The Matched Line 157
7. Approximate Values of γ. Propagation Velocity as a Function of Frequency (Dispersion) 158
8. Impedance Seen Looking into a Line 160
9. The Lossless Line as an Impedance Transformer 163
10. Conclusion 166

Chapter VI. **Solutions to Exercises**

Exercises for Chapter II 169
Exercises for Chapter III 179
Exercises for Chapter IV 189

SUBJECT INDEX 207

Notation

Representation of the line (two wire)

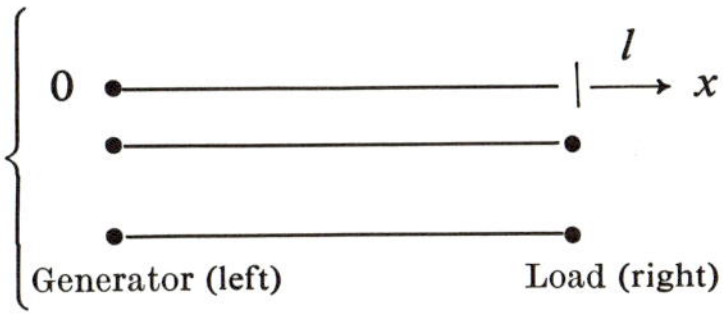

Length of the line	l
Distance along the line	x
Time	t
Frequency	f
Radian frequency	$\omega = 2\pi f$
Input subscript ($x = 0$)	$0 \left(\text{Ex} : \mathrm{Z}_0 = \dfrac{\mathrm{V}_0}{\mathrm{I}_0}\right)$
Output subscript ($x = l$)	$l \left(\text{Ex} : \mathrm{Z}_l = \dfrac{\mathrm{V}_l}{\mathrm{I}_l}\right)$
Generator subscript ($x = 0$)	0
Load subscript ($x = l$)	l
Self-inductance per unit length	L
Capacitance per unit length	C
Series resistance per unit length	R
Parallel conductance per unit length	G
Impedance per unit length, variable p	$\mathrm{Z}(p) = \mathrm{R} + \mathrm{L}p$
Impedance per unit length, variable ω	$\mathrm{Z}(\omega) = \mathrm{R} + j\mathrm{L}\omega$
Characteristic impedance, variable p	$\mathrm{Z}_c = \left(\dfrac{\mathrm{R} + \mathrm{L}p}{\mathrm{G} + \mathrm{C}p}\right)^{1/2}$
Characteristic impedance, variable ω	$\mathrm{Z}_c = \left(\dfrac{\mathrm{R} + j\mathrm{L}\omega}{\mathrm{G} + j\mathrm{C}\omega}\right)^{1/2}$
Characteristic resistance	$\mathrm{R}_c = (\mathrm{L}/\mathrm{C})^{1/2}$

Propagation function { variable p	$\gamma(p) = [(R + Lp)(G + Cp)]^{1/2}$
Propagation function { variable ω	$\gamma(\omega) = \alpha(\omega) + j\beta(\omega) = [(R + jL\omega)(G + jC\omega)]^{1/2}$
Attenuation (in nepers per meter)	α
Speed of light	c
Propagation velocity	u
Propagation constant, or delay per unit length	$\delta = \sqrt{LC} = \left[\dfrac{1}{u_{\text{lossless}}}\right]$
Propagation time from one end of the line to the other	$\tau = \delta l$
Wavelength	λ
Internal impedance of the generator feeding the line	Z_0
Load impedance connected to the output of the line	Z_l
Reflection coefficient at the input	$\Gamma_0 = \dfrac{Z_0 - Z_c}{Z_0 + Z_c}$
Reflection coefficient at the output	$\Gamma_l = \dfrac{Z_l - Z_c}{Z_l + Z_c}$
A complex number	A
Absolute value of a complex number	$\lvert A \rvert$
Voltage at a point with abscissa x { variable p	$V(x, p)$
Voltage at a point with abscissa x { variable t	$v(x, t)$
Voltage at a point with abscissa x { variable ω	$V = \mathscr{V}(x)e^{j\omega t}$
Current at a point with abscissa x { variable p	$I(x, p)$
Current at a point with abscissa x { variable t	$i(x, t)$
Current at a point with abscissa x { variable ω	$I = \mathscr{I}(x)e^{j\omega t}$
Voltage and current amplitudes in the sinusoidal steady state	$\mathscr{V}(x),\ \mathscr{I}(x)$
Time constant of the first-order system associated with the line	T
Penetration depth (skin effect)	d
Time constant due to skin effect	$T_{0.5}$
Laplace transform pair	$f(t) \Leftrightarrow F(p)$
Unit Step function	$\Upsilon(t)$

Transmission Lines with Pulse Excitation

CHAPTER I

General Equations for Transmission Lines

1. Introduction

1.1 *The first fundamental hypothesis—homogeneity*

Homogeneous transmission lines consist of an assemblage of at least two conductors, arranged parallel to an axis $x'x$ (Fig. 1). The geometric and physical parameters of the line (the nature of the conductors and of the

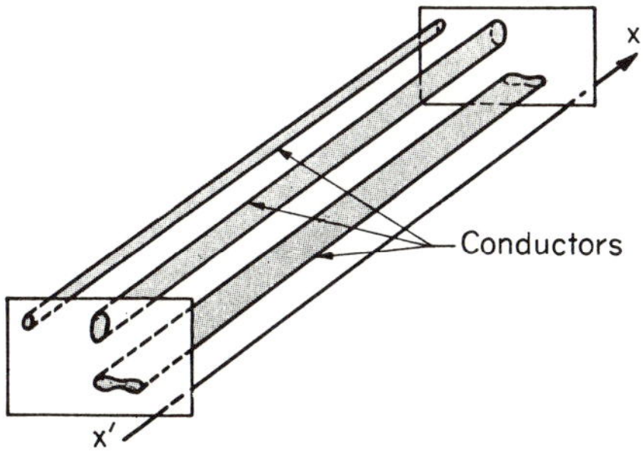

Fig. 1. A general transmission line.

dielectric) are assumed to be constant everywhere along the line; this is the hypothesis of homogeneity of the line. This assemblage of conductors comprises two groups of at least one conductor each, one group being called the *forward conductors*, and the other the *return conductors*.

1.2 *The second fundamental hypothesis—conservation of current*

The simplified theory of lines that is to be treated here assumes that the lateral dimensions of the line are negligible, or, more precisely, that the time of propagation of the electromagnetic field between the forward and return conductors in a plane perpendicular to the axis of the line is negligible with respect to the duration of the briefest of the phenomena to be studied. This restriction leads to the second fundamental hypothesis, that of the *conservation of current* across a plane perpendicular to the line. The total current crossing this plane is zero, which is to say that the current through the forward conductors is equal, but in the opposite sense, to the current through the return conductors.

This second hypothesis of the simplified theory is in fact a consequence of Maxwell's equations, which rest on quite general hypotheses.†

These two fundamental hypotheses reduce the theory of transmission lines to a problem in partial differential equations in two variables (time and one space variable taken along the axis of the line). The general case would lead to partial differential equations in four variables (time and three space variables).

1.3 *Remarks*

(a) With no loss of generality, it can be assumed that the line is composed of only two conductors. This considerably simplifies the definition of per unit length parameters of the line, and, in fact, most actual lines are of this type.

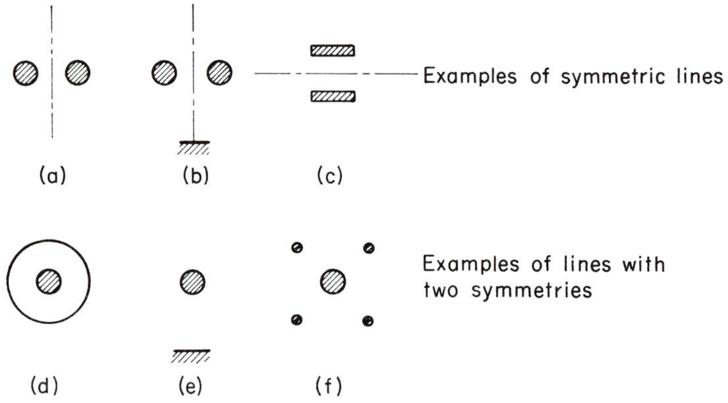

Fig. 2. Some particular line configurations.

† On this subject, the book by E. Roubine, "Lignes et Antennes," p. 43 (*Revue d'Optique, Paris* **1**, 1954), may be consulted, among others.

(b) Frequently the additional hypothesis of *symmetry* of the line is made, motivated by the two-wire line having two identical cylindrical conductors. This assumption refers to the symmetry of the line with respect to a plane parallel to its axis (Fig. 2a–c). This hypothesis is by no means necessary, since the single-wire line above a conducting plane and the coaxial line are not excluded from the simplified theory, but in certain cases it can simplify the development. We will appeal to this simplification in establishing the general line equations.

2. Definition of the State of the Line at a Point

Let us consider an ideal two-wire transmission line, having forward and return conductors reduced to straight lines (Fig. 3). Let the line MN be the intersection of the transmission line with a plane perpendicular to the

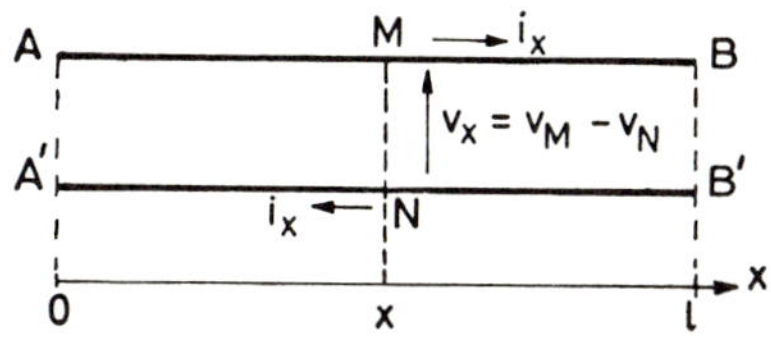

Fig. 3. A two-wire transmission line.

line and located a distance x along an axis Ox parallel to the line. Arbitrarily choose AB to be the forward conductor and A′B′ to be the return conductor.

2.1 *Current*

The *current* in the line at the abscissa x is defined to be the current flowing in conductor AB at point M. It is taken to be positive if it is directed in the sense Ox, i.e., if it flows from M towards B. By the hypothesis of conservation of current, the current at N is equal and opposite to the current at M.

2.2 *Voltage*

If v_M and v_N are, respectively, the potentials of points M and N with respect to some reference potential, the *line voltage* at the abscissa x will be

$$v_x = v_M - v_N$$

3. Per Unit Length Parameters, or Primary Parameters

3.1 *Remark*

We will use the term *parameter* for the quantities to be defined below, rather than the term *constant* used by most authors. The latter term arises from the fact that discussions of transmission lines most often consider only sinusoidal waves of a *given* frequency. On the contrary, we will be especially interested in the pulse regime, which corresponds to a large *band* of frequencies. At least two of the parameters of interest are strong functions of frequency: the resistance per unit length, affected by the skin effect, and the conductance per unit length, affected by the variations of the dielectric losses with frequency. Nevertheless, in deriving the equations of interest, and in solving them using the methods of Chapters II–IV, we will suppose that these parameters are constants.

3.2 *Resistance per unit length*

The line being homogeneous, the total resistance of the conductors is proportional to the length of the line. A *resistance per unit length* R can thus be defined.

3.3 *Self-inductance per unit length*

In the same way a *self-inductance per unit length* can be defined, which is the result of the true self-inductance of the conductors and the mutual inductance between the two conductors. The self-inductance per unit length will be denoted by L.

3.4 *Conductance per unit length*

Because of imperfect insulation between the conductors (losses in the dielectric which separates them), a uniformly distributed transverse conductance appears. It is thus possible to define a transverse *conductance per unit length* G.

3.5 *Capacitance per unit length*

This is the ratio of the charge on a unit length conductor element to the voltage between the two conductors at the element considered. The *capacitance per unit length* is denoted C.

4. General Equations

4.1 *The general case*

Consider a line element of length Δx (Fig. 4). This element can be compared to a quadripole (AA′, BB′) with elements

$$\mathrm{R}\,\Delta x, \quad \mathrm{L}\,\Delta x, \quad \mathrm{G}\,\Delta x, \quad \text{and } \mathrm{C}\,\Delta x$$

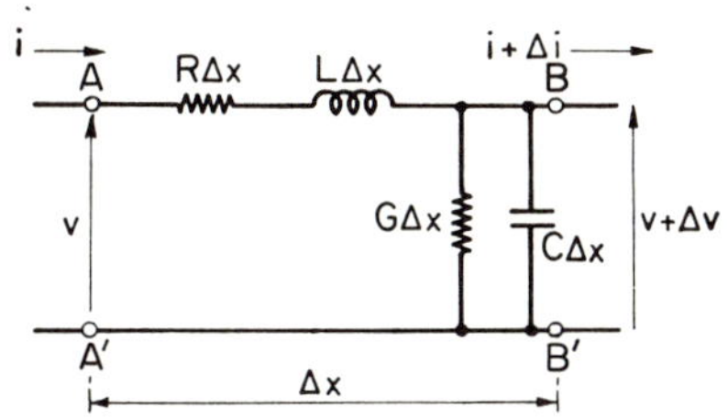

Fig. 4. An infinitesimal line element.

The voltage rise from A to B will be

$$\Delta v = v_{\mathrm{B}} - v_{\mathrm{A}} = -\mathrm{R}\,\Delta x\, i - \mathrm{L}\,\Delta x \frac{\partial i}{\partial t} \tag{I.1}$$

The current Δi flowing into B from B′ is

$$\Delta i = -\mathrm{G}\,\Delta x\,(v + \Delta v) - \mathrm{C}\,\Delta x \frac{\partial}{\partial t}(v + \Delta v)$$

$$\Delta i = -\mathrm{G}\,\Delta x\, v - \mathrm{C}\,\Delta x \frac{\partial v}{\partial t} + \mathrm{G}\,\Delta x^2 \left(\mathrm{R}i + \mathrm{L}\frac{\partial i}{\partial t}\right) + \mathrm{C}\,\Delta x^2 \left(\mathrm{R}\frac{\partial i}{\partial t} + \mathrm{L}\frac{\partial^2 i}{\partial t^2}\right) \tag{I.2}$$

Dividing the terms of (I.1) by Δx and letting Δx tend to 0, we obtain the *first fundamental equation*:

$$\lim_{\Delta x \to 0}\left(\frac{\Delta v}{\Delta x}\right) = \frac{\partial v}{\partial x} = -\mathrm{R}i - \mathrm{L}\frac{\partial i}{\partial t} \tag{I.3}$$

In the same way, from (I.2),

$$\lim_{\Delta x \to 0}\left(\frac{\Delta i}{\Delta x}\right) = \lim_{\Delta x \to 0}\left\{-\mathrm{G}v - \mathrm{C}\frac{\partial v}{\partial t} + \Delta x\left[\mathrm{G}\left(\mathrm{R}i + \mathrm{L}\frac{\partial i}{\partial t}\right) + \mathrm{C}\left(\mathrm{R}\frac{\partial i}{\partial t} + \mathrm{L}\frac{\partial^2 i}{\partial t^2}\right)\right]\right\}$$

which results in the *second fundamental equation*:

$$\frac{\partial i}{\partial x} = -\mathrm{G}v - \mathrm{C}\frac{\partial v}{\partial t} \tag{I.4}$$

Differentiating (I.3) with respect to x yields

$$\frac{\partial^2 v}{\partial x^2} = -\left[\mathrm{R}\frac{\partial i}{\partial x} + \mathrm{L}\frac{\partial}{\partial x}\left(\frac{\partial i}{\partial t}\right)\right] = -\left[\mathrm{R}\frac{\partial i}{\partial x} + \mathrm{L}\frac{\partial}{\partial t}\left(\frac{\partial i}{\partial x}\right)\right]$$

Replacing $\partial i/\partial x$ in this by its value in (I.4) leads to

$$\begin{aligned}\frac{\partial^2 v}{\partial x^2} &= -\mathrm{R}\left(-\mathrm{G}v - \mathrm{C}\frac{\partial v}{\partial t}\right) - \mathrm{L}\frac{\partial}{\partial t}\left(-\mathrm{G}v - \mathrm{C}\frac{\partial v}{\partial t}\right)\\ &= \mathrm{RG}v + \mathrm{RC}\frac{\partial v}{\partial t} + \mathrm{LG}\frac{\partial v}{\partial t} + \mathrm{LC}\frac{\partial^2 v}{\partial t^2}\end{aligned}$$

from which, finally,

$$\frac{\partial^2 v}{\partial x^2} = \mathrm{RG}v + (\mathrm{RC} + \mathrm{LG})\frac{\partial v}{\partial t} + \mathrm{LC}\frac{\partial^2 v}{\partial t^2} \tag{I.5}$$

This last relation is the *telegraphers' equation.* It can only be integrated in certain special cases, e.g., if the voltage v is sinusoidal (see Chapter V), or in the transient regime using operational calculus (see Section 5).

The current equation is of the same form as (I.5), and can be obtained by differentiating (I.4) with respect to x.

4.2 *The special case of the lossless line*

If the parameters G and R can be neglected, the fundamental equations (I.3) and (I.4) simplify to

$$\frac{\partial v}{\partial x} = -\mathrm{L}\frac{\partial i}{\partial t} \tag{I.6}$$

$$\frac{\partial i}{\partial x} = -\mathrm{C}\frac{\partial v}{\partial t} \tag{I.7}$$

These relations lead to the equation for the propagation of plane waves, which can also be obtained from (I.5) by setting $\mathrm{R} = 0$, $\mathrm{G} = 0$:

$$\frac{\partial^2 v}{\partial x^2} = \mathrm{LC}\frac{\partial^2 v}{\partial t^2} \tag{I.8}$$

Setting

$$u = \frac{1}{(\mathrm{LC})^{1/2}} = \frac{1}{\delta} \tag{I.9}$$

in which u is the *propagation velocity* and δ is the *propagation constant*, or the delay per unit length, this equation takes the form

$$\frac{\partial^2 v}{\partial x^2} - \frac{1}{u^2} \cdot \frac{\partial^2 v}{\partial t^2} = 0 \tag{I.10}$$

The interpretation of u and δ will be given in Chapters II and IV.

The solution of (I.10) is of the form

$$v(x, t) = g(x - ut) + h(x + ut) \tag{I.11}$$

in which g and h are arbitrary functions. The current in the line can now be found from relations (I.6), (I.7), and (I.11), which yield

$$i(x, t) = \frac{1}{R_c} [g(x - ut) - h(x + ut)] \tag{I.12}$$

where

$$R_c = (L/C)^{1/2} \tag{I.13}$$

is by definition the *characteristic resistance* of the line.

5. Use of Operational Calculus (The Laplace Transform)†

5.1 *The propagation function*

Consider a transmission line such as defined above, having per unit parameters L, C, R, and G, and of length l, as shown in Fig. 5.

At the instant $t = 0$, an electromotive force $e(t)$, arising from a generator with internal impedance Z_0, is applied to the left end, or input, of the line. The right end, or output, of the line is terminated in an impedance Z_l.

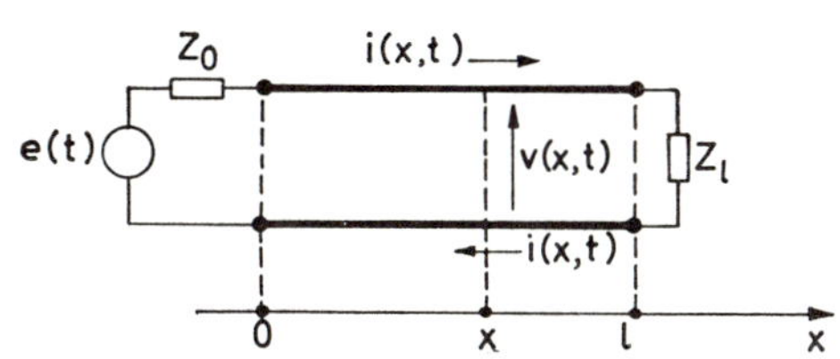

Fig. 5. The transmission line to be studied.

† For background on the Laplace transformation, the reader can refer to Chapter I of the book by Metzger and Vabre, "Electronique des Impulsions," Vol. 1, *Circuits à Constantes Localisées*, Masson, Paris, or to numerous other texts.

The currents and voltages at each point along the line are assumed to be zero prior to the initial time $t = 0$. The problem is to calculate the voltage $v(x, t)$ between the two conductors of the line, and the current $i(x, t)$ flowing in each of these conductors, at each point x and at each instant t. The distance x is taken positive from the input of the line as origin towards the output of the line.

In Section 4, two relations between the voltage $v(x, t)$ and the current $i(x, t)$ were established:

$$\frac{\partial v(x, t)}{\partial x} + \mathrm{R}\, i(x, t) + \mathrm{L} \frac{\partial i(x, t)}{\partial t} = 0 \tag{I.3}$$

$$\frac{\partial i(x, t)}{\partial x} + \mathrm{G}\, v(x, t) + \mathrm{C} \frac{\partial v(x, t)}{\partial t} = 0 \tag{I.4}$$

Let us rewrite these equations in terms of the Laplace transforms of the current and voltage functions. To that end consider

$$\mathrm{V}(x, p) \Leftrightarrow v(x, t); \qquad \mathrm{I}(x, p) \Leftrightarrow i(x, t)$$

It is well known that

$$\frac{\partial v(x, t)}{\partial t} \Leftrightarrow p\, \mathrm{V}(x, p) - v(x, 0^+); \qquad \frac{\partial i(x, t)}{\partial t} \Leftrightarrow p\, \mathrm{I}(x, p) - i(x, 0^+)$$

Then the voltage and current in the line being assumed zero before the initial instant, and the voltage and current not being able to change in an infinitesimal time, the quantities $v(x, 0^+)$ and $i(x, 0^+)$ are zero for all x along the line. Hence

$$\frac{\partial v(x, t)}{\partial t} \Leftrightarrow p\, \mathrm{V}(x, p); \qquad \frac{\partial i(x, t)}{\partial t} \Leftrightarrow p\, \mathrm{I}(x, p)$$

Since

$$\mathrm{V}(x, p) = \int_0^{+\infty} e^{-pt}\, v(x, t)\, dt$$

we have

$$\frac{\partial \mathrm{V}(x, p)}{\partial x} = \int_0^{+\infty} e^{-pt} \frac{\partial v(x, t)}{\partial x}\, dt \Leftrightarrow \frac{\partial v(x, t)}{\partial x}$$

and (I.3) and (I.4) in the transform domain become

$$\frac{\partial \mathrm{V}(x, p)}{\partial x} + (\mathrm{R} + \mathrm{L}p)\, \mathrm{I}(x, p) = 0 \tag{I.14}$$

$$\frac{\partial \mathrm{I}(x, p)}{\partial x} + (\mathrm{G} + \mathrm{C}p)\, \mathrm{V}(x, p) = 0 \tag{I.15}$$

Differentiating (I.14) with respect to x and using (I.15) to eliminate $\partial \mathrm{I}(x, p)/\partial x$ from the result, we obtain

$$\frac{\partial^2 \mathrm{V}(x, p)}{\partial x^2} - (\mathrm{R} + \mathrm{L}p)(\mathrm{G} + \mathrm{C}p)\, \mathrm{V}(x, p) = 0$$

An analogous relation can be found for the current:

$$\frac{\partial^2 \mathrm{I}(x, p)}{\partial x^2} - (\mathrm{R} + \mathrm{L}p)(\mathrm{G} + \mathrm{C}p)\, \mathrm{I}(x, p) = 0$$

Let us introduce

$$\gamma(p) = [(\mathrm{R} + \mathrm{L}p)(\mathrm{G} + \mathrm{C}p)]^{1/2} \tag{I.16}$$

This is called the *propagation function.* Its physical significance will be discussed in Chapter IV. To simplify the notation, $\gamma(p)$ will be written simply as γ. Note that γ is independent of x, but not of p. We then have finally

$$\frac{\partial^2 \mathrm{V}(x, p)}{\partial x^2} - \gamma^2\, \mathrm{V}(x, p) = 0 \tag{I.17}$$

$$\frac{\partial^2 \mathrm{I}(x, p)}{\partial x^2} - \gamma^2\, \mathrm{I}(x, p) = 0 \tag{I.18}$$

5.2 *Characteristic impedance*

The differential equation (I.17) has solutions of the form

$$\mathrm{V}(x, p) = \mathrm{V}_1(p)\, e^{-\gamma x} + \mathrm{V}_2(p)\, e^{\gamma x} \tag{I.19}$$

where $\mathrm{V}_1(p)$ and $\mathrm{V}_2(p)$ are arbitrary functions of p only, which will be written simply as V_1 and V_2.

From (I.19) there follows

$$\frac{\partial \mathrm{V}(x, p)}{\partial x} = -\gamma(\mathrm{V}_1\, e^{-\gamma x} - \mathrm{V}_2\, e^{\gamma x})$$

which together with (I.14) yields

$$\mathrm{I}(x, p) = \frac{\gamma}{\mathrm{R} + \mathrm{L}p}(\mathrm{V}_1\, e^{-\gamma x} - \mathrm{V}_2\, e^{\gamma x}) = \left(\frac{\mathrm{G} + \mathrm{C}p}{\mathrm{R} + \mathrm{L}p}\right)^{1/2}(\mathrm{V}_1\, e^{-\gamma x} - \mathrm{V}_2\, e^{\gamma x})$$

It is conventional to define

$$\left(\frac{\mathrm{R} + \mathrm{L}p}{\mathrm{G} + \mathrm{C}p}\right)^{1/2} = \mathrm{Z}_c(p) \tag{I.20}$$

This last quantity, which has the dimensions of an impedance, is called the *characteristic impedance* of the line. It is related to the physical properties of the line, i.e., to its dimensions and to the properties of the materials used in its construction. It is a function of p, and hence of time. We will write $Z_c(p)$ simply as Z_c.

The general solutions of (I.17) and (I.18) are thus

$$V(x, p) = V_1 e^{-\gamma x} + V_2 e^{\gamma x} \tag{I.21}$$

$$I(x, p) = \frac{V_1 e^{-\gamma x} - V_2 e^{\gamma x}}{Z_c} \tag{I.22}$$

5.3 *Reflection coefficients*

The complete solution to this problem is obtained by determining the functions V_1 and V_2 using the conditions at the ends of the line.

Let $E(p) \Leftrightarrow e(t)$ be the transform of the generator voltage. At the input to the line $(x = 0)$

$$E(p) = Z_0 I(0, p) + V(0, p) = \frac{Z_0}{Z_c}(V_1 - V_2) + V_1 + V_2$$

using (I.21) and (I.22), or

$$V_1\left(\frac{Z_0}{Z_c} + 1\right) - V_2\left(\frac{Z_0}{Z_c} - 1\right) = E(p)$$

or again

$$V_1 - V_2\frac{Z_0 - Z_c}{Z_0 + Z_c} = E(p)\frac{Z_c}{Z_0 + Z_c}$$

Let us define

$$\Gamma_0 = \frac{Z_0 - Z_c}{Z_0 + Z_c} \tag{I.23}$$

This is called the *voltage reflection coefficient* at the *input* to the line. Its physical significance will become apparent in Chapters II and IV. A simple calculation shows that $|\Gamma_0|$ is always between 0 and 1. Using this quantity, we have

$$V_1 - \Gamma_0 V_2 = E(p)\frac{Z_c}{Z_0 + Z_c} \tag{I.24}$$

In the same way, at the output of the line $(x = l)$,

$$Z_l = \frac{V(l, p)}{I(l, p)} = Z_c\frac{V_1 e^{-\gamma l} + V_2 e^{\gamma l}}{V_1 e^{-\gamma l} - V_2 e^{\gamma l}}$$

or

$$\frac{Z_l}{Z_c}(V_1 e^{-\gamma l} - V_2 e^{\gamma l}) = V_1 e^{-\gamma l} + V_2 e^{\gamma l}$$

$$V_1 e^{-\gamma l}\left(\frac{Z_l}{Z_c} - 1\right) - V_2 e^{\gamma l}\left(\frac{Z_l}{Z_c} + 1\right) = 0$$

or finally

$$V_1 e^{-\gamma l}\frac{Z_l - Z_c}{Z_l + Z_c} - V_2 e^{\gamma l} = 0$$

Let us define now

$$\Gamma_l = \frac{Z_l - Z_c}{Z_l + Z_c} \tag{I.25}$$

This is the *voltage reflection coefficient* at the *output* of the line. It is easily seen that $|\Gamma_l|$ is always between 0 and 1 also. Using this quantity,

$$\Gamma_l V_1 e^{-\gamma l} - V_2 e^{\gamma l} = 0 \tag{I.26}$$

Thus for the determination of V_1 and V_2 we have the two equations

$$V_1 - \Gamma_0 V_2 = E(p)\frac{Z_c}{Z_0 + Z_c} \tag{I.24}$$

$$\Gamma_l V_1 e^{-\gamma l} - V_2 e^{\gamma l} = 0 \tag{I.26}$$

Solving these yields

$$V_1 = E(p)\frac{Z_c}{Z_0 + Z_c}\,\frac{e^{\gamma l}}{e^{\gamma l} - \Gamma_0\Gamma_l e^{-\gamma l}} = E(p)\frac{Z_c}{Z_0 + Z_c}\,\frac{1}{1 - \Gamma_0\Gamma_l e^{-2\gamma l}}$$

$$V_2 = \Gamma_l V_1 e^{-2\gamma l} = E(p)\,\frac{Z_c}{Z_0 + Z_c}\,\frac{\Gamma_l e^{-2\gamma l}}{1 - \Gamma_0\Gamma_l e^{-2\gamma l}}$$

which when substituted into (I.21) and (I.22) result in

$$V(x, p) = E(p)\frac{Z_c}{Z_0 + Z_c}\,\frac{e^{-\gamma x} + \Gamma_l e^{-\gamma(2l-x)}}{1 - \Gamma_0\Gamma_l e^{-2\gamma l}} \tag{I.27}$$

$$I(x, p) = E(p)\frac{1}{Z_0 + Z_c}\,\frac{e^{-\gamma x} - \Gamma_l e^{-\gamma(2l-x)}}{1 - \Gamma_0\Gamma_l e^{-2\gamma l}} \tag{I.28}$$

When Γ_0 and Γ_l, which are functions of p, are both different from zero, it is difficult to compute the inverse transforms of $V(x, p)$ and $I(x, p)$. As an aid in this, the following series expansion can be used:

$$\frac{1}{1 - \Gamma_0\Gamma_l e^{-2\gamma l}} = 1 + \Gamma_0\Gamma_l e^{-2\gamma l} + \cdots + \Gamma_0^n\Gamma_l^n e^{-2n\gamma l} + \cdots$$

$$= \sum_{n=0}^{\infty} \Gamma_0^n\Gamma_l^n e^{-2n\gamma l}$$

This substitution can be justified by noting that the series on the right is a geometric series with argument $\Gamma_0\Gamma_l\, e^{-2\gamma l}$, the absolute value of which is less than one. The series thus converges. Using this series in (I.27) and (I.28) yields the following basic relations:

5.4 *Basic relations*

$$V(x, p) = E(p)\frac{Z_c}{Z_0 + Z_c}\,[e^{-\gamma x} + \Gamma_l\, e^{-\gamma(2l-x)}] \sum_{n=0}^{\infty} \Gamma_0^n \Gamma_l^n\, e^{-2n\gamma l} \qquad (I.29)$$

$$I(x, p) = E(p)\frac{1}{Z_0 + Z_c}\,[e^{-\gamma x} - \Gamma_l\, e^{-\gamma(2l-x)}] \sum_{n=0}^{\infty} \Gamma_0^n \Gamma_l^n\, e^{-2n\gamma l} \qquad (I.30)$$

with

$$\gamma = [(R + Lp)(G + Cp)]^{1/2} \qquad (I.16)$$

$$Z_c = \left(\frac{R + Lp}{G + Cp}\right)^{1/2} \qquad (I.20)$$

$$\Gamma_0 = \frac{Z_0 - Z_c}{Z_0 + Z_c} \qquad (I.23)$$

$$\Gamma_l = \frac{Z_l - Z_c}{Z_l + Z_c} \qquad (I.25)$$

6. Appendix. The Per Unit Length Parameters L, C, R, G

6.1 *The electrical phenomena giving rise to the parameters*

These four parameters depend on the geometry of the conductors and insulation, and on the electric, dielectric, and magnetic properties of the materials used in constructing the transmission line. The geometric factors are independent of frequency, but the physical properties (resistivity, dielectric constant, permeability) are in general functions of frequency. For simplicity, we will study the electrical phenomena only at a fixed frequency, rather than over a range of frequencies as would be present in the pulse regime.

1. *Skin effect*

In the case of direct current in a single isolated conductor, the current density is uniform across the conductor. In the case of alternating current, however, the current density is not uniform. The current appears to be concentrated in a "skin," of greater or lesser thickness, on the surface of the conductor. The thickness of this skin is

$$d = \frac{1}{(\pi f \mu \sigma)^{1/2}}$$

(see the Appendix to Chapter IV), where d is in meters, f is the frequency in hertz, $\mu = \mu_0 \mu_r$ is the magnetic permeability of the conducting material in henries per meter, and σ is the conductivity of the conducting material in (ohm-meter)$^{-1}$. For copper,

$$\sigma = 5.85 \times 10^7 \quad (\Omega\text{meter})^{-1}$$

$$\mu = 4\pi \times 10^{-7} \quad (\text{H/meter})$$

which yields, for example,

f	50 Hz	10 kHz	1 MHz	100 MHz
d (copper)	9.2 mm	0.65 mm	0.065 mm	6.5 μ

Remark. In actual fact, the current density decreases gradually from the surface of the conductor towards the center (Fig. 6). The use of d

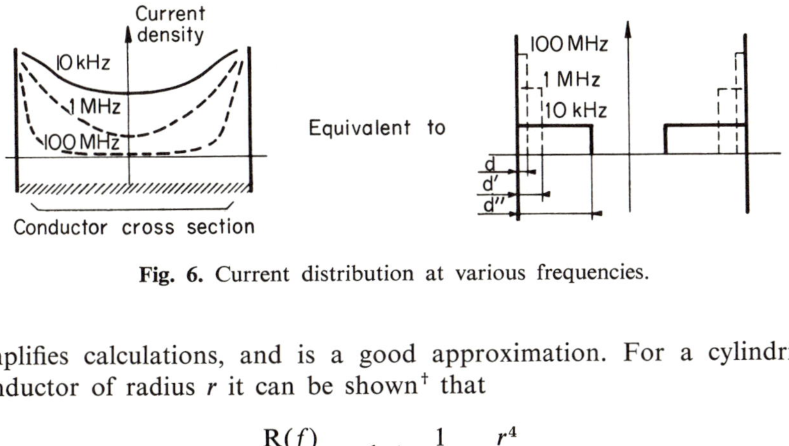

Fig. 6. Current distribution at various frequencies.

simplifies calculations, and is a good approximation. For a cylindrical conductor of radius r it can be shown† that

$$\frac{\mathrm{R}(f)}{\mathrm{R}_{\mathrm{D.C.}}} = 1 + \frac{1}{48} \cdot \frac{r^4}{d^4}$$

where

$$\mathrm{R}_{\mathrm{D.C.}} = \varrho \frac{l}{\pi r^2}$$

† S. Ramo and J. Whinnery, "Fields and Waves in Modern Radio." Wiley, New York, 1953; P. Grivet and R. Legros, "Physique des Circuits." Masson, Paris, 1960.

2. *The effect of the proximity of neighboring conductors*

The currents flowing through two neighboring conductors are subject to mutual repulsion (or attraction) governed by the laws of electromagnetism. The current density in each conductor is thus less (or greater) than otherwise at points near the other conductor. The density is affected in the opposite sense at points farthest from the other conductor. The result is an apparent decrease in the cross section of the conductors, and correspondingly an apparent increase in the resistance.

For two identical conductors of radius r, having centers separated a distance D, it can be shown† that the increase ΔR in the resistance is

$$\frac{\Delta R}{R_{D.C.}} = \frac{G(Q)}{1 - \frac{4r^2}{D^2} H(Q)} \cdot \frac{4r^2}{D^2}$$

where $Q = 2\pi r(2\mu\sigma f)^{1/2}$ is a dimensionless quantity, and G(Q) and H(Q) are the monotonically increasing functions of Q shown in Fig. 7.

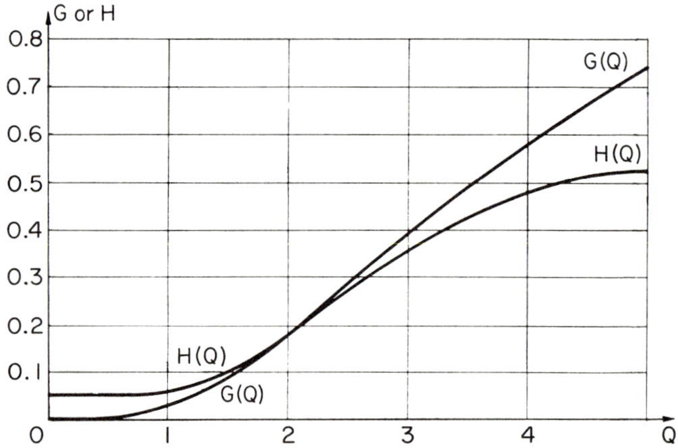

Fig. 7. Relative to the proximity effect.

As an example, for a two-wire line, whether or not twisted, with conductors 1 mm in diameter and having axes separated 2 mm, at 200 kHz,

$$\frac{\Delta R}{R_{D.C.}} = 0.22$$

That is, the *proximity effect* increases the resistance 22% above the direct current value.

† R. Croze and L. Simon, "Transmissions Téléphoniques." Eyrolles, Paris, 1956.

3. *The effect of radiation*

The currents flowing in the line induce eddy currents in neighboring metallic (conducting or magnetic) masses. These eddy currents dissipate energy as heat or in magnetizing the magnetic materials. This energy loss, to the detriment of the inducing currents, causes an apparent increase in the resistance of the line, which is greater the higher the frequency. This effect can be prevented by the use of coaxial cable.

4. *Dielectric loss*

Since a perfect dielectric does not exist, it is necessary to associate a conductivity with any real dielectric. This conductivity generally depends on frequency through an experimental law which must be determined for each dielectric material. In practice, for materials such as Teflon, this loss is negligible at frequencies below 100 MHz, and even up to 1000 MHz this effect is negligible in comparison to the loss due to skin effect.

6.2 *Influence of these phenomena on the parameters*

1. *Resistance per unit length*

To the direct current resistance, there is added a term which is a function of frequency and which arises from the skin effect, the proximity effect, and the effects of radiation. At high frequencies (above 100 MHz, i.e., for pulses with rise time less than 10 nsec), skin effect is usually preponderant, since in this case coaxial lines are usually used, for which radiation and proximity effects are not present. For open-wire lines, all three effects must be considered. The proximity effect is important for closely spaced conductors, such as a twisted two-wire line. The radiation effects depend on what the surroundings of the open-wire line are. It can be measured, if necessary, but it is seldom possible to calculate it in practice.

2. *The self-inductance per unit length*

To the low-frequency self-inductance (constant current density) there is added a term which again depends on the skin effect, the proximity effect, and the effects of radiation. The first two factors affect the geometric dimensions of the current tubes (these differ from the conductors in which they flow). The last factor creates a mutual inductance between the inducing current and the induced currents, which modifies the self-inductance of the primary, or inducing, circuit (self-inductance transferred by the transformer effect).

These factors have been discussed by Terman,* and by Croze and Simon.†

* F. Terman, "Radio Engineers Handbook." McGraw-Hill, New York, 1943.

† R. Croze and L. Simon. "Transmissions Téléphoniques." Eyrolles, Paris, 1956.

First example. Consider a two-wire line with identical conductors separated by a distance large with respect to their diameter (Fig. 8). The self-inductance per unit length of the two conductors at low frequency is

$$L = 0.92 \times 10^{-6} \log_{10}\left(\frac{D}{r}\right) \quad \text{(H/meter)}$$

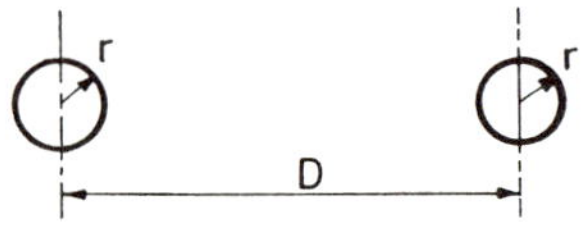

Fig. 8. A thin-conductor transmission line.

For each conductor, skin effect adds an inductance†

$$L_i = \frac{3.16 \times 10^{-4}}{2\pi r}\left(\frac{\mu_r}{\sigma f}\right)^{1/2} \approx \frac{10^{-4}}{2r}\left(\frac{\mu_r}{\sigma f}\right)^{1/2} \quad \text{(H/meter)}$$

where μ_r is the relative permeability, f is the frequency in hertz, and σ is the conductivity. Thus the total self-inductance per unit length is

$$L + 2L_i = 0.92 \times 10^{-6} \log_{10}\left(\frac{D}{r}\right) + \frac{10^{-4}}{r}\left(\frac{\mu_r}{\sigma f}\right)^{1/2}$$

Second example. Consider a two-wire line with very closely spaced conductors (Fig. 9). The proximity effect arises, and the current density

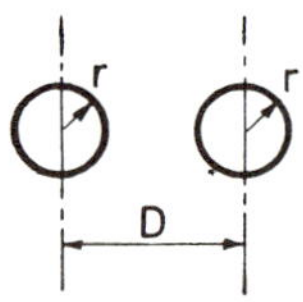

Fig. 9. A thick-conductor transmission line.

is higher in the parts of the conductors nearest to one another. A fictitious equivalent radius of the conductors can be calculated to be

$$r' = r\left(1 - \left(\frac{2r}{D}\right)^2\right)^{1/2}$$

† This formula results from the approximate relation

$$\frac{L_i\omega}{R_{D.C.}} = \frac{1}{2}\,\frac{r}{d}\,\frac{1}{l}$$

See the book by S. Ramo and J. Whinnery (Sections 6–10, p. 248) cited p. 13.

and the fictitious equivalent separation is

$$D' = \frac{D}{2}\left(1 + \left(1 - \left(\frac{2r}{D}\right)^2\right)^{1/2}\right)$$

Then for each conductor, the inductance added due to skin effect is

$$L_i = \frac{3.16 \times 10^{-4}}{2\pi r'}\left(\frac{\mu_r}{\sigma f}\right)^{1/2} = \frac{10^{-4}}{2r\left(1 - \left(\frac{2r}{D}\right)^2\right)^{1/2}}\left(\frac{\mu_r}{\sigma f}\right)^{1/2}$$

The low-frequency self-inductance is

$$L = 0.92 \times 10^{-6} \log_{10}\left(\frac{D'}{r}\right) = 0.92 \times 10^{-6} \log_{10}\left[\frac{D\left(1 + \left(1 - \left(\frac{2r}{D}\right)^2\right)^{1/2}\right)}{2r}\right]$$

and the total self-inductance per unit length is

$$2L_i + L$$

Third example. A coaxial line (Fig. 10). The low-frequency self-inductance per unit length (henry/meter) is

$$L = 0.46 \times 10^{-6} \log_{10}\left(\frac{r_2}{r_1}\right)$$

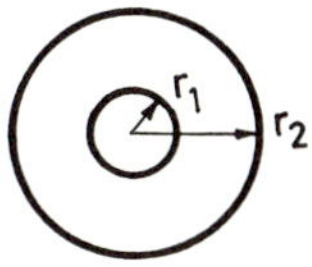

Fig. 10. A coaxial line. Geometry.

The self-inductance per unit length added by skin effect is for conductors 1 and 2

$$L_{i_1} = \frac{10^{-4}}{2r_1}\left(\frac{\mu_r}{\sigma f}\right)^{1/2} \quad \text{(H/meter)}$$

$$L_{i_2} = \frac{10^{-4}}{2r_2}\left(\frac{\mu_r}{\sigma f}\right)^{1/2} \quad \text{(H/meter)}$$

The total self-inductance per unit length is

$$L + L_{i_1} + L_{i_2}$$

3. *Conductance* G *per unit length*

This is usually negligible with respect to the resistance per unit length R. When this is not the case, the relation

$$G = C\omega \tan \phi$$

can be used, where ω is the radian frequency of interest and $\tan\phi$ is a coefficient characteristic of the dielectric. The angle ϕ is called the *loss angle.*

4. *Capacitance* C *per unit length*

This depends only on the geometry of the conductors and the dielectric constant of the insulating material between the conductors. Various formulas for this will be given below.

6.3 *Formulas for the values of the per unit length parameters*

The units of the International System (MKS system) will be used.

1. *The constants of vacuum, or free space*

Dielectric constant:

$$\varepsilon_0 = \frac{10^{-9}}{36\pi} \quad \text{(F/meter)}$$

Absolute permeability:

$$\mu_0 = 4\pi \times 10^{-7} \quad \text{(H/meter)}$$

Speed of light:

$$c = \frac{1}{(\varepsilon_0\mu_0)^{1/2}} = 3 \times 10^8 \quad \text{meter/sec}$$

Wave impedance:

$$\eta_0 = \left(\frac{\mu_0}{\varepsilon_0}\right)^{1/2} = 120\pi = 377\,\Omega$$

2. *Constants of the material used*

Relative permeability:	μ_r
Absolute permeability:	$\mu = \mu_0\mu_r$
Relative dielectric constant:	ε_r
Absolute dielectric constant:	$\varepsilon = \varepsilon_0\varepsilon_r$
Speed of propagation:	$u = c/(\varepsilon_r\mu_r)^{1/2}$
Resistivity (ohm-meter):	ϱ
Tangent of the loss angle:	$\tan\phi$

3. *The coaxial line*

The dielectric is assumed nonmagnetic, $\mu_r = 1$, and the two conductors are assumed to have the same resistivity ϱ (Fig. 11).

$$(a) \quad \mathrm{L} = \frac{\mu}{2\pi}\,\mathrm{Log}\,\frac{d_2}{d_1} = 2\times 10^{-7}\,\mathrm{Log}\,\frac{d_2}{d_1} = 4.6\times 10^{-7}\,\log\frac{d_2}{d_1} \quad \text{(H/meter)}$$

(Here Log is the natural logarithm, base e, and log is the common logarithm, base 10.)

$$(b)\quad \mathrm{R} = \left(\frac{\mu f \varrho}{\pi}\right)^{1/2} \frac{d_1 + d_2}{d_1 d_2} = 0.632 \times 10^{-3} (\varrho f)^{1/2} \frac{d_1 + d_2}{d_1 d_2} \qquad (\Omega/\text{meter})$$

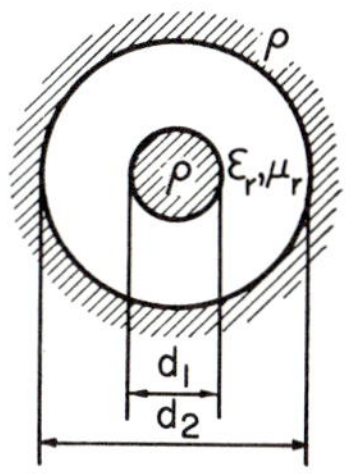

Fig. 11. A coaxial line. Parameters.

In the case of copper conductors,

$$\mathrm{R} = 8.3 \times 10^{-8} f^{1/2} \frac{d_1 + d_2}{d_1 d_2} \qquad (\Omega/\text{meter})$$

$$(c)\quad \mathrm{C} = \frac{2\pi\varepsilon}{\operatorname{Log} \dfrac{d_2}{d_1}} = \frac{10^{-9}}{18} \times \frac{\varepsilon_r}{\operatorname{Log} \dfrac{d_2}{d_1}} = 0.24 \times 10^{-10} \frac{\varepsilon_r}{\log \dfrac{d_2}{d_1}} \qquad (\text{F/meter})$$

$$(d)\quad \mathrm{G} = \mathrm{C}\omega \tan \varphi = 2\pi f \mathrm{C} \tan \varphi \qquad (\Omega^{-1}/\text{meter})$$

$$(e)\quad \mathrm{R}_c = \left(\frac{\mathrm{L}}{\mathrm{C}}\right)^{1/2} = \frac{1}{2\pi} \left(\frac{\mu}{\varepsilon}\right)^{1/2} \operatorname{Log} \frac{d_2}{d_1} = \frac{1}{2\pi} \eta_0 \left(\frac{\mu_r}{\varepsilon_r}\right)^{1/2} \operatorname{Log} \frac{d_2}{d_1}$$

$$\mathrm{R}_c = \frac{60}{\varepsilon_r^{1/2}} \operatorname{Log} \frac{d_2}{d_1} = \frac{138}{\varepsilon_r^{1/2}} \log \frac{d_2}{d_1} \qquad (\Omega)$$

4. *Various other lines*

The formulas in Fig. 12 use the MKS units. The dielectric is assumed to be air. For other homogeneous dielectrics, divide R_c by $\varepsilon_r^{1/2}$. Note that L and C are obtained simply by dividing R_c and $1/\mathrm{R}_c$, respectively, by 3×10^8. The parameter G is

$$\frac{2.09 \times 10^{-8}}{\mathrm{R}_c} \varepsilon_r^{1/2} (\tan \varphi) f$$

In the case of thin wires, R can be found quickly by attributing to each

	Characteristic resistance (Ω)	Capacitance per unit length (pF/meter)	Self-inductance per unit length (μH/meter)
Asymmetric two-wire line 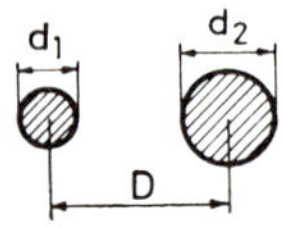	$R_c = 120 \operatorname{Log} \frac{2D}{\sqrt{d_1 d_2}}$	$C = \frac{27.5}{\operatorname{Log} \frac{2D}{\sqrt{d_1 d_2}}}$	$L = 0.4 \operatorname{Log} \frac{2D}{\sqrt{d_1 d_2}}$
Symmetric two-wire line 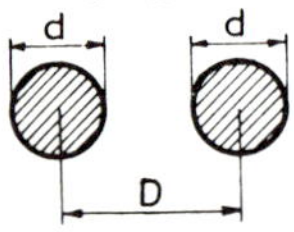	$R_c = 120 \operatorname{Arc\,cosh} \frac{D}{d}$ $= 120 \operatorname{Log} \frac{2D}{d}$	$C = \frac{10^3}{36 \operatorname{Arc\,cosh} \frac{D}{d}}$ $= \frac{27.5}{\operatorname{Log} \frac{2D}{d}}$	$L = 0.4 \operatorname{Arc\,cosh} \frac{D}{d}$ $= 0.4 \operatorname{Log} \frac{2D}{d}$
Shielded two-wire line 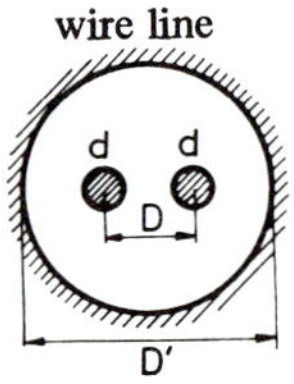	$R_c = 120 \operatorname{Log}\left(\frac{2D}{d} \frac{D'^2-D^2}{D'^2+D^2}\right)$	$C = \frac{27.5}{\operatorname{Log}\left(\frac{2D}{d} \frac{D'^2-D^2}{D'^2+D^2}\right)}$	$L = 0.4 \operatorname{Log}\left(\frac{2D}{d} \frac{D'^2-D^2}{D'^2+D^2}\right)$
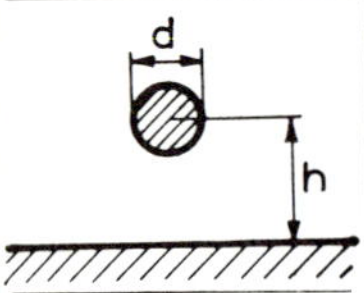	$R_c = 60 \operatorname{Log} \frac{4h}{d}$	$C = \frac{55}{\operatorname{Log} \frac{4h}{d}}$	$L = 0.2 \operatorname{Log} \frac{4h}{d}$
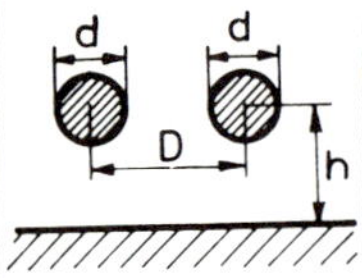	$R_c = 120 \operatorname{Log} \frac{2D}{d\sqrt{1+\left(\frac{D}{2h}\right)^2}}$	$C = \frac{27.5}{\operatorname{Log} \frac{2D}{d\sqrt{1+\left(\frac{D}{2h}\right)^2}}}$	$L = 0.4 \operatorname{Log} \frac{2D}{d\sqrt{1+\left(\frac{D}{2h}\right)^2}}$

Fig. 12. Parameters for various line configurations.

wire of diameter d the resistance per unit length:

$$0.63 \times 10^{-3} \frac{1}{(\varrho f)^{1/2}} \frac{1}{d}$$

A more complete set of formulas can be found in the book by Roubine.†

† E. Roubine, "Lignes et antennes," *Revue d'Optique, Paris* **1**, 1954.

CHAPTER II

The Method of Traveling Waves

1. The Method of Traveling Waves for the Lossless Line

1.1 *Decomposition into two traveling waves*

In Chapter I, Section 4.2, it was shown that the state at a point along the line with abscissa x, at time t, could be described in terms of two relations giving the voltage and the current:

$$v(x, t) = g(x - ut) + h(x + ut) \qquad \text{with} \quad u = \frac{1}{(\mathrm{LC})^{1/2}} = \frac{1}{\delta}$$

$$i(x, t) = \frac{g(x - ut)}{\mathrm{R}_c} - \frac{h(x + ut)}{\mathrm{R}_c}$$

in which u has the dimensions of a velocity, and R_c those of a resistance. It will now be shown that each of the two pairs of functions

$$\mathrm{V}_i = g(x - ut) \qquad \mathrm{I}_i = \frac{1}{\mathrm{R}_c} g(x - ut) = \frac{\mathrm{V}_i}{\mathrm{R}_c}$$

and

$$\mathrm{V}_r = h(x + ut) \qquad \mathrm{I}_r = \frac{-1}{\mathrm{R}_c} h(x + ut) = \frac{-\mathrm{V}_r}{\mathrm{R}_c}$$

describes a traveling wave on the line. These two waves propagate with the same speed in directions opposite to one another.

Consider the term V_i. In the (x, t) plane, the curves of equal phase $x - ut = \text{const}$, or $dx/dt - u = 0$, or $dx/dt = u$, are parallel straight lines with slope u (Fig. 13). The phase $(x - ut)$, and hence $\mathrm{V}_i = g(x - ut)$, has the same value at time t_2 and abscissa x_2, as it had at some earlier time t_1 at abscissa x_1. Thus V_i represents a traveling wave, propagating in the positive direction (from A towards B) with a constant velocity u. To the voltage wave V_i there corresponds a current wave V_i/R_c which also propagates from A to B with the same velocity u.

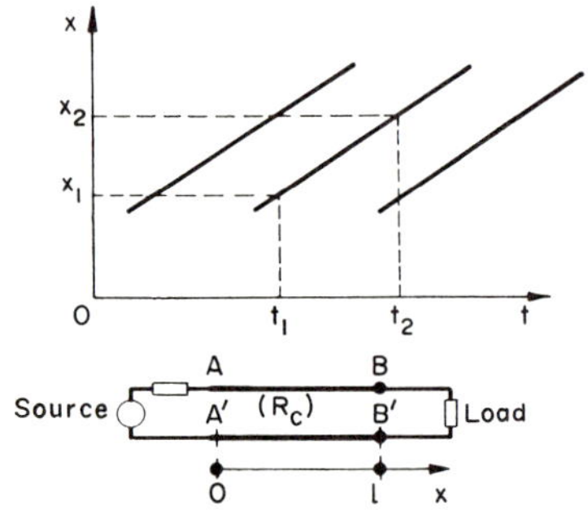

Fig. 13. Equiphase contours for incident waves.

The same reasoning applied to $V_r = h(x + ut)$ and $I_r = -V_r/R_c$ shows that V_r represents a wave traveling in the negative direction (from B towards A) with constant speed u, which is to say that the velocity of the wave is $-u$. To the voltage V_r corresponds a current wave $-V_r/R_c$ also propagating from B towards A with the same speed u. This current has the opposite sign to the voltage V_r, but the directions of propagation of V_r and I_r are the same.

1.2 *The incident and reflected waves*

Usually energy flows from A towards B, so that it is conventional to call (V_i, I_i) the *incident wave* and (V_r, I_r) the *reflected wave*. It will be shown in Section 5.2 of the present chapter, and in Section 2.7 of Chapter IV, that in the case of pulsed operation the voltage (and current) at an arbitrary point along the line is the superposition of voltage (current) waves, each having the same form as the voltage (current) pulse applied at the input, but having different amplitudes, and appearing at successive instants in time. These times differ by multiples of the time τ of propagation through the line.

Thus the incident voltage wave V_i can be decomposed into the sum of waves v_i, v_i', v_i'', ... appearing at times θ, $\theta + 2\tau$, $\theta + 4\tau$, ..., where θ is the propagation time from the input to the point x of interest. In the same way, the reflected wave V_r can be decomposed into the sum of waves v_r, v_r', v_r'', ... appearing at times $2\tau - \theta$, $4\tau - \theta$, $6\tau - \theta$, ... The same is true for the current waves.

1.3 *The characteristic resistance* R_c

In Section 1.1 this was seen to be the ratio between the voltage and current amplitudes of the same traveling wave, to within the sign. More precisely,

$$\frac{v_i}{i_i} = R_c; \qquad \frac{v_r}{i_r} = -R_c$$

Thus in its propagation, each voltage wave sees an impedance R_c.

1.4 *Reflection coefficients*

These will be defined taking $Z_0 = R_0$ and $Z_l = R_l$ to be real, in contrast to Section 5.3 of Chapter I.

(a) The Output Reflection coefficient (Fig. 14). Whatever positive sense is chosen for the currents, Kirchhoff's laws applied to end B of the line yield

$$i_i + i_r = i = \text{current in } R_l$$

$$v_i + v_r = R_l i = \text{voltage between B and B}'$$

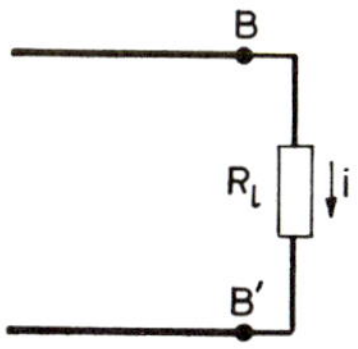

Fig. 14. The load terminating a line.

From these,

$$\frac{v_i}{R_c} - \frac{v_r}{R_c} = \frac{v_i + v_r}{R_l}$$

$$v_i\left(\frac{1}{R_c} - \frac{1}{R_l}\right) = v_r\left(\frac{1}{R_c} + \frac{1}{R_l}\right)$$

$$v_r = \frac{R_l - R_c}{R_l + R_c} v_i$$

The quantity

$$\Gamma_l = \frac{R_l - R_c}{R_l + R_c}$$

is defined to be the *voltage reflection coefficient* at the end of the line, or at the *load*.

In the same way, for the currents it can be shown that

$$i_r = -\frac{R_l - R_c}{R_l + R_c}\, i_i$$

Thus

$$v_r = \Gamma_l v_i; \qquad i_r = -\Gamma_l i_i \tag{II.1}$$

These also hold true if v_i, v_r, i_i, i_r are replaced, respectively, by V_i, V_r, I_i, I_r.

(b) Reflection Coefficient at the Input (Fig. 15 with $Z_0 = R_0$, $Z_l = R_l$). Starting from the beginning of the excitation and over a duration 2τ, only the incident wave affects the input end of the line (see Section 1.2). In that case,

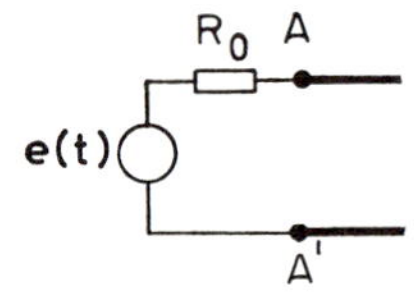

Fig. 15. A generator feeding a line.

$$v_i = e(t)\frac{R_c}{R_0 + R_c}; \qquad i_i = \frac{e(t)}{R_0 + R_c}$$

At the instant 2τ, a wave (v_r, i_r) reflected from the load arrives at the input, is again reflected, and gives rise to a second incident wave (v_i', i_i'). Then

$$\begin{aligned} v_i + v_i' + v_r &= e(t) - R_0(i_i + i_i' + i_r) \\ &= e(t) - \frac{R_0}{R_c}(v_i + v_i' - v_r) \end{aligned}$$

from which, using the expressions for v_i and i_i,

$$v_i' = \frac{R_0 - R_c}{R_0 + R_c} v_r; \qquad i_i' = -\frac{R_0 - R_c}{R_0 + R_c}\, i$$

Defining

$$\Gamma_0 = \frac{R_0 - R_c}{R_0 + R_c}$$

to be the *reflection coefficient* at the *input* to the line, these become

$$v_i' = \Gamma_0 v_r; \qquad i_i' = -\Gamma_0 i_r \tag{II.2}$$

The coefficients Γ_0 and Γ_l vary from -1 to $+1$.

1.5 *Summary of formulas for the method of traveling waves*

For simplicity, it is usual to assume that the incident (and then the reflected) wave represents the *change* of the voltage or current from that which existed before the pulse. Thus to the two waves, incident and reflected, the initial values of the voltage and current are added:

$$\begin{aligned} v(x, t) &= \mathrm{V}_i(x, t) + \mathrm{V}_r(x, t) + v_0(x), \qquad & \mathrm{V}_i &= \mathrm{R}_c \mathrm{I}_i \\ i(x, t) &= \mathrm{I}_i(x, t) + \mathrm{I}_r(x, t) + i_0(x), \qquad & \mathrm{V}_r &= -\mathrm{R}_c \mathrm{I}_r \end{aligned} \tag{II.3}$$

$$\mathrm{R}_c = (\mathrm{L}/\mathrm{C})^{1/2}$$

These formulas are valid for signals of arbitrary waveform.

In addition, for a line of length l, a parameter τ is defined, which is the time of propagation of the incident wave from the input to the output of the line, which is also the time of propagation of the reflected wave from the output to the input, since they have the same propagation velocity:

$$\tau = \frac{l}{u} = l(\mathrm{LC})^{1/2}$$

2. Principal Results

2.1 *Various loads*

(a) MATCHED LOAD (Fig. 16). It was shown in Section 1.3 that each voltage wave sees an impedance R_c as it propagates. It is easy to see that if the incident wave encounters an impedance R_c at the output end, it is not reflected. For we have $v_r = \Gamma_l v_i$. But if $\mathrm{R}_l = \mathrm{R}_c$, then $\Gamma_l = 0$, so that

$$v_r = 0; \qquad i_r = 0$$

$$\mathrm{V_B} = v_i + v_0; \qquad \mathrm{I_B} = i_i + i_0$$

The impedance R_c which so terminates the line can be either lumped, or distributed (a transmission line with the same R_c). It thus results as a special case that the impedance seen looking into a semi-infinite line at any point is R_c.

(b) THE OPEN-CIRCUITED LINE (Fig. 17). Here we have

$$R_l = \infty; \qquad \Gamma_l = +1$$

so that

$$v_i = v_r; \qquad i_i = -i_r$$

There is thus *total reflection*, *without* change of sign for the voltage, and *with* change of sign for the current, so that

$$V_B = 2v_i + v_0; \qquad I_B = 0$$

(c) THE SHORT-CIRCUITED LINE (Fig. 18). Here

$$R_l = 0; \qquad \Gamma_l = -1$$

so that

$$v_i = -v_r; \qquad i_i = i_r$$

Thus there is again total reflection, but *with* change of sign for the voltage, and *without* change of sign for the current, so that

$$V_B = 0; \qquad I_B = 2i_i + i_0$$

2.2 *Various generators*

(a) THE MATCHED GENERATOR (Fig. 19). Here $v_i' = \Gamma_0 v_r$, but $R_0 = R_c$ so that $\Gamma_0 = 0$, and hence $v_i' = 0$. Thus the wave returning from the load is not reflected at the input. Hence the incident wave is independent of the reflected wave, and remains equal for all time to (v_i, i_i), the incident wave on the interval 0 to 2τ. Thus

$$V_A = v_i + v_r + v_0 \qquad \text{with} \quad v_i = \frac{U(t) - v_0 - R_c i_0}{2}$$

$$I_A = i_i + i_r + i_0 \qquad \text{with} \quad i_i = \frac{U(t) - v_0 - R_c i_0}{2R_c}$$

(b) CURRENT GENERATOR (Fig. 20). Here

$$I_A = I(t); \qquad R_0 = \infty; \qquad \Gamma_0 = +1$$

from which

$$v_i' = v_r; \qquad i_i' = -i_r$$

Thus the wave returning from the output is totally reflected, without change of sign for the voltage, and with change of sign for the current.

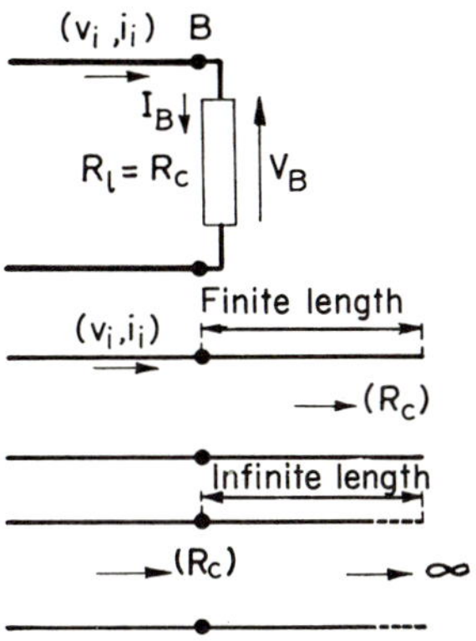

Fig. 16. A matched load.

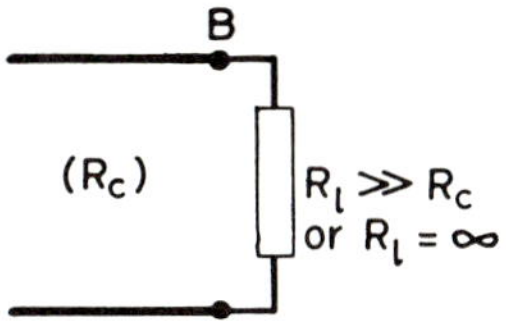

Fig. 17. The open-circuited line.

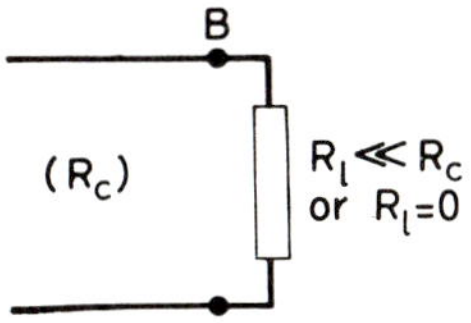

Fig. 18. The short-circuited line.

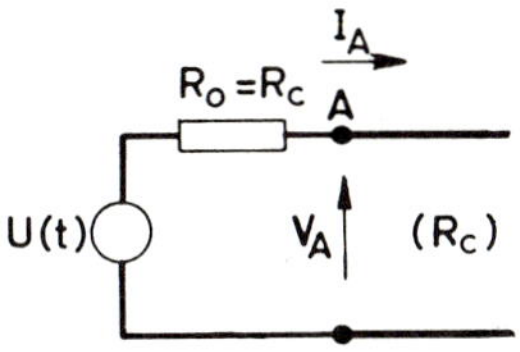

Fig. 19. Matched source feeding a line.

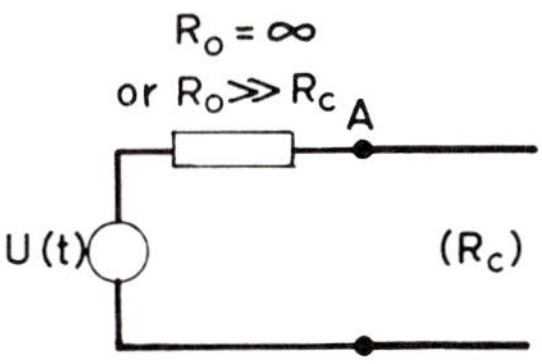

Fig. 20. Current generator feeding a line.

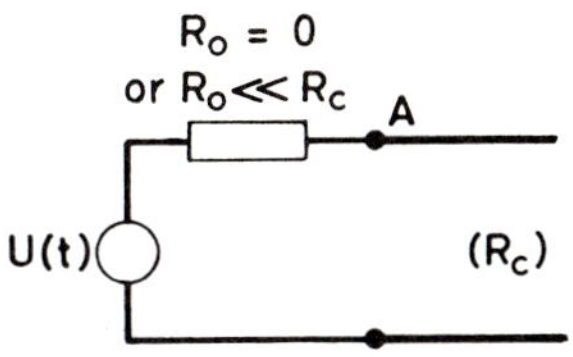

Fig. 21. Voltage generator feeding the line

(c) VOLTAGE GENERATOR (Fig. 21). Here

$$V_A = U(t); \qquad R_0 = 0; \qquad \Gamma_0 = -1$$

from which

$$v_i' = -v_r; \qquad i_i' = i_r$$

Thus the wave returning from the load is again totally reflected, but with change of sign for the voltage, and without change of sign for the current.

2.3 *The general case*

In general, there are *partial reflections*, with a decrease in the amplitude of each successive wave. According to the equations of Sections 1.4(a) and 1.4(b), at the output,

$$v_r = \Gamma_l v_i; \qquad i_r = -\Gamma_l i_i$$

At the input, starting from a time 2τ after the beginning of the excitation, the reflection of the reflected wave (produced at the load) is

$$v_i' = \Gamma_0 v_r; \qquad i_i' = -\Gamma_0 i_r$$

The new incident wave is then $(v_i' + v_i)$, $(i_i' + i_i)$. This new incident wave is in its turn reflected at the output with reflection coefficient Γ_l, then returns to the input, etc.

Note that if the voltage source $U(t)$ is variable with time, the reflected wave which reaches the input at an instant t is a fraction ($v_r = \Gamma_l v_i$) of the incident wave which issued from the source at an instant 2τ earlier, i.e., at time $t - 2\tau$. This latter incident wave is not necessarily the same as that which issues from the source at the time t.

Note: To find the voltage and current at the load, with a matched source, the line can be replaced by a voltage source $2v_i$ in series with a resistance R_c, provided the initial conditions are zero ($v_0 = i_0 = 0$). Then

$$i_l = \frac{2v_i}{R_l + R_c}; \qquad v_l = \frac{2v_i R_l}{R + R_c}$$

This is true since

$$i_l = i_i + i_r = i_i(1 - \Gamma_l) = \frac{v_i}{R_c}\left(1 - \frac{R_l - R_c}{R_l + R_c}\right)$$

$$i_l = \frac{v_i}{R_c} \cdot \frac{2R_c}{R_l + R_c}$$

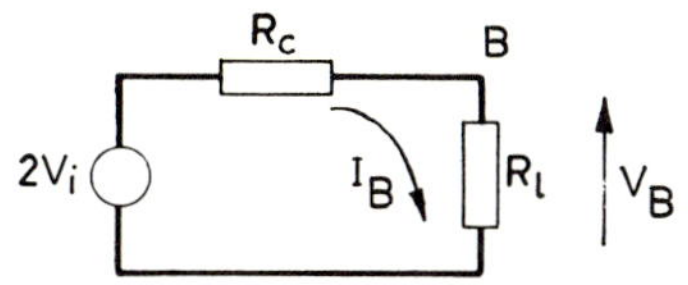

Fig. 22. Equivalent circuit for a line with matched source.

If $V_i = \varepsilon v_i$, $I_B = \varepsilon i_l$, $V_B = \varepsilon v_l$ (since $v_0 = 0$, $i_0 = 0$) then

$$I_B = \frac{2V_i}{R_l + R_c}$$

$$V_B = \frac{2V_i\, R_l}{R_l + R_c}$$

from which follows the equivalent circuit of Fig. 22.

3. Examples

Note: In this paragraph, R_0 represents a lumped resistance connected at the input to the line. In calculating Γ_0, to R_0 should be added the internal impedance of the generator.

3.1 *Charging a line using* $R_0 = 0$, $R_l = \infty$[†]

1. *The transistor rise time is* $\ll \tau$ *(the propagation time)*

(a) Statement of the Problem (Fig. 23). Consider a line charged by a saturated transistor, producing a voltage step. In this example and the

† There are two systems of circuit representation of transistors. The first symbol below is that usually used in Europe, the second is that usually used in America:

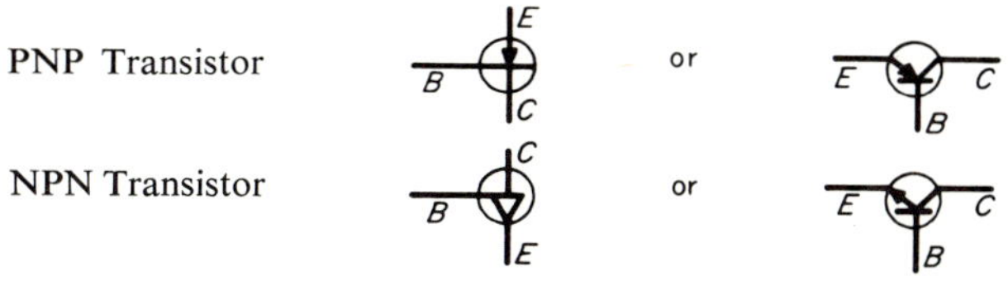

following one, the voltage drops in the transistor and in the clamping diode will be neglected.

(b) LOSSLESS LINE Consider first the voltage (Fig. 23). At point A, from time 0 to 2τ the incident wave is $v_i = +U$, $v_r = 0$. Since V_A is zero

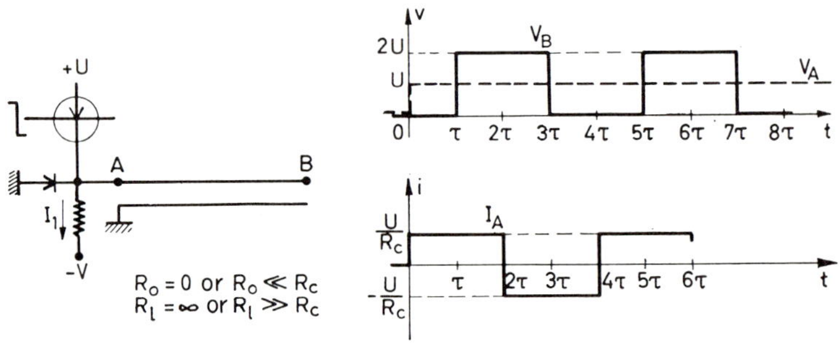

Fig. 23. Transistor charging an open line.

for $t < 0$, then $V_A = 0 + v_i = +U$ from 0 to 2τ. At B, from τ to 3τ, v_i is reflected without change of sign, so that

$$v_r = v_i = +U; \qquad V_B = v_i + v_r = 2U$$

At A, from 2τ to 4τ, v_r is reflected as from a voltage generator, so that

$$v_i' = -v_r = -U; \qquad V_A = v_i + v_r + v_i' = +U$$

and the voltage at A remains at $+U$. At B, from 3τ to 5τ, v_i' is reflected without change of sign, so that

$$v_r' = v_i' = -U; \qquad V_B = v_i + v_r + v_i' + v_r' = 0$$

etc.

For the current (Fig. 23), from 0 to 2τ,

$$I_A = \frac{v_i}{R_c} = \frac{U}{R_c} \quad (v_r = 0)$$

from 2τ to 4τ,

$$I_A = i_i + i_r + i_i' = \frac{v_i}{R_c} - \frac{v_r}{R_c} + \frac{v_i'}{R_c}$$

$$I_A = \frac{U}{R_c} - \frac{U}{R_c} - \frac{U}{R_c} = -\frac{U}{R_c}$$

etc.

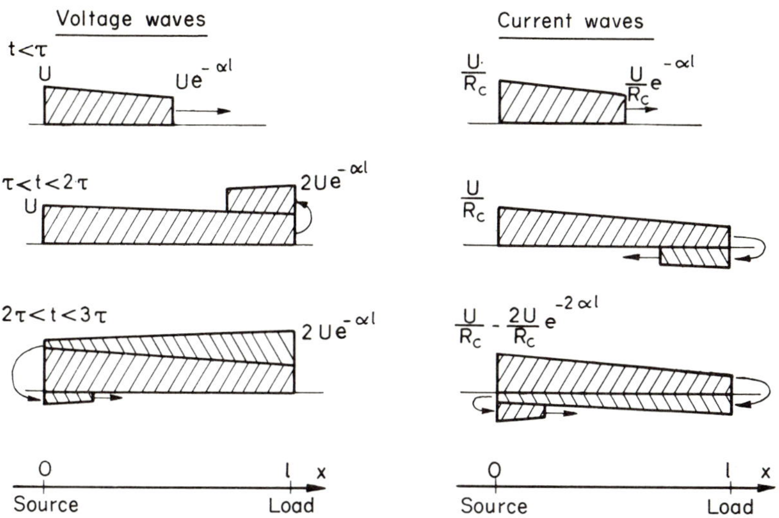

Fig. 24. Waves on a lossy line.

(c) LINE WITH LOSS (Figs. 24 and 25). In the Appendix to this chapter it will be shown that losses affect $\gamma(p)$ (see Chapter I, Section 5.1) in such a way that, for most cases (lines without dispersion, lines at the beginning of the transient regime), it is true that $\gamma(p) = p\delta + \alpha$. Since this coefficient enters into the various relations of interest only as an exponent, it is possible

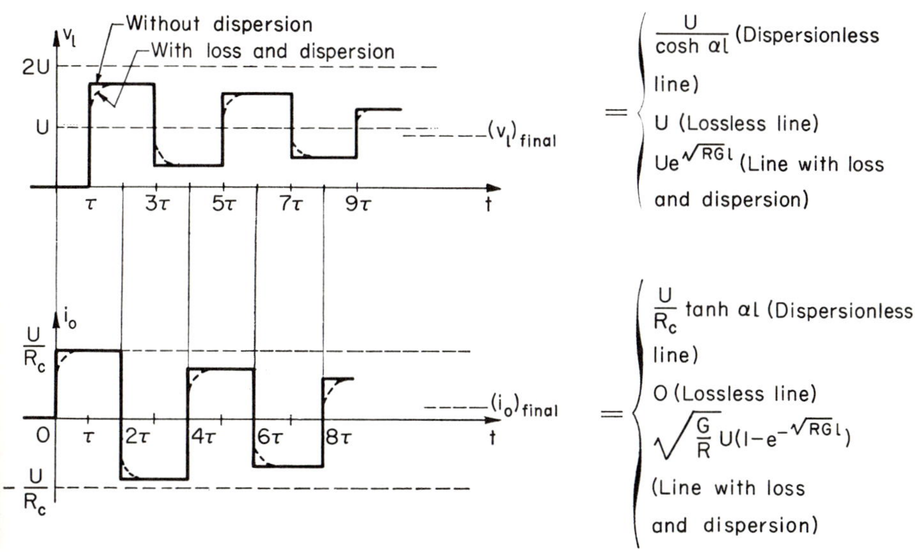

Fig. 25. Voltage and current on a lossy line.

to separate $p\delta$ and α into factors as $e^{-p\delta x}e^{-\alpha x}$. The factor $e^{-p\delta x}$ represents a delay of δx, and the factor $e^{-\alpha x}$ represents a decrease in the amplitude of the wave.

On leaving A, the incident wave v_i equals $+U$. When it arrives at B, it has decayed to $Ue^{-\alpha l}$. Taking account of this decrease, the reasoning of the preceding paragraphs can be gone through again, and it will be seen that the voltage V_B appears as a sequence of decreasing values. Two cases will be considered:

(1) Line with Loss but without Dispersion of Velocities. These are the lines which satisfy the Heaviside condition. Although this case is of no practical interest, it is simpler than the case of a real line, and allows the effects of loss and of dispersion of velocities to be separated. For this type of line, the waveforms remain rectangular, but the amplitudes of successive pulses decrease. By adding the amplitudes of successive waves, the asymptotic wave amplitudes can be obtained:

$$2U \sum_{K=0}^{\infty} (-1)^K e^{-(2K+1)\alpha l} = 2Ue^{-\alpha l}[1 - e^{-2\alpha l} + \cdots + (-1)^K e^{-2K\alpha l} + \cdots]$$

$$= \frac{2Ue^{-\alpha l}}{1 + e^{-2\alpha l}} = \frac{U}{\cosh(\alpha l)}$$

for the output voltage, and

$$\frac{U}{R_c} - \frac{2U}{R_c}\sum_{K=0}^{\infty}(-1)^K e^{-(2K+2)\alpha l} = \frac{U}{R_c}\,\frac{1 - e^{-2\alpha l}}{1 + e^{-2\alpha l}} = \frac{U}{R_c}\tanh(\alpha l)$$

for the input current. These show that in the steady state, as expected, there is a voltage drop along the line, so that v_l (final) $< U$, as well as leakage currents, so that i_0 (final) $\neq 0$.

(2) Line with Both Loss and Dispersion of Velocities. An actual line has both loss and dispersion of the velocities of propagation of various frequencies. In this case, the pulses decrease in amplitude as before, but are no longer rectangular, since the propagation velocities of the various elementary waves in the voltage (or current) spectrum are not the same.

The above formulas are no longer applicable. Returning to the basic equations of Chapter I, there results

$$\frac{\partial v}{\partial x} = -Ri - L\frac{\partial i}{\partial t}$$

$$\frac{\partial i}{\partial x} = -Gv - C\frac{\partial v}{\partial t}$$

In the steady state, $\partial v/\partial t = \partial i/\partial t = 0$ and L and C no longer have an effect. There remain only

$$\left.\begin{aligned} \frac{\partial v}{\partial x} &= -\mathrm{R}i \\ \frac{\partial i}{\partial x} &= -\mathrm{G}v \end{aligned}\right\} \Rightarrow \frac{\partial^2 v}{\partial x^2} = \mathrm{RG}v$$

from which

$$v(x, t = \infty) = \mathrm{M} \exp[+(\mathrm{RG})^{1/2}x] + \mathrm{N} \exp[-(\mathrm{RG})^{1/2}x]$$

Setting $x = \infty$ in this last shows that we must have $\mathrm{M} = 0$ (the voltage can not increase indefinitely, even though the length of the line becomes longer and longer). Then setting $x = 0$ shows that $\mathrm{N} = \mathrm{U}$. Thus

$$v(l, t = \infty) = \mathrm{U} \exp[-(\mathrm{RG})^{1/2}l]$$

which is the asymptotic value of $\mathrm{V_B}$.

In the same way, the current at A oscillates between positive and negative values. A calculation similar to the above shows that the asymptotic value is

$$i(0, t = \infty) = \mathrm{U}(\mathrm{G}/\mathrm{R})^{1/2}(1 - \exp[-(\mathrm{RG})^{1/2}l])$$

2. *The transistor collector current is a ramp (rise time $> \tau$)*

(a) STATEMENT OF THE PROBLEM (Fig. 26). In the first stage, the collector current is a ramp $i = kt$ (we will assume $k = \mathrm{I_1}/\tau$). Until i exceeds $\mathrm{I_1}$, the diode is cut off, and $\mathrm{V_A}$ can increase (this instant will be taken as origin). In the second stage, $\mathrm{V_A}$ holds fixed at the value $+\mathrm{U}$.

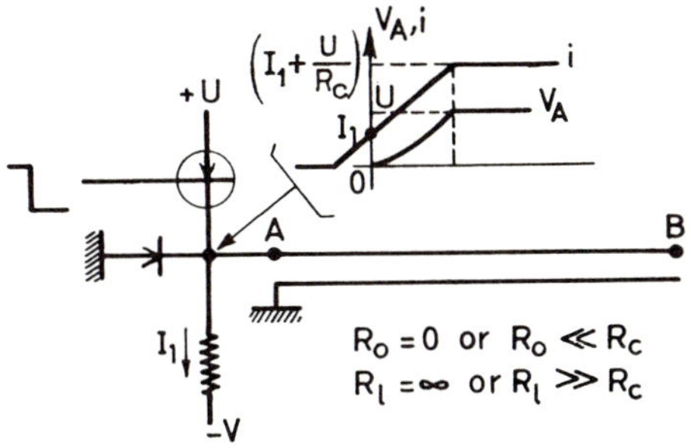

Fig. 26. Line charged by a current ramp.

(b) FIRST STAGE. Here the transistor is operating normally as a current generator. The reflection coefficient $\Gamma_0 = 1$ since the impedance at the input of the line is infinite (see the note at the beginning of Section 3).

At A, from 0 to 2τ,

$$I_A = i_i = kt; \qquad V_A = v_i = R_c i_i = kR_c t$$

and the voltage V_A increases linearly with slope kR_c.

At B, from τ to 3τ, v_i is totally reflected without change of sign, so that $v_r = v_i$. But v_i is the incident wave at the input shifted by the transmission time τ, $v_i = kR_c(t - \tau)$, so that $V_B = v_i + v_r = 2kR_c(t - \tau)$. Thus V_B increases with slope $2kR_c$.

At A, from 2τ to 4τ, v_r is reflected totally without change of sign (current generator), $v_i' = v_r$. But v_r is the incident wave, reflected at the output, and shifted in time by the roundtrip propagation time 2τ. Thus

$$v_i' = v_r = kR_c(t - 2\tau)$$

$$V_A = v_i + v_r + v_i' = kR_c t + 2kR_c(t - 2\tau)$$

$$V_A = kR_c(3t - 4\tau)$$

and V_A increases with slope $3kR_c$.

In the same way, from 3τ to 5τ the voltage at B increases linearly with slope $4kR_c$, and from 4τ to 6τ the voltage at A increases with slope $5kR_c$, etc. (Fig. 27).

(c) SECOND STAGE. Here the transistor is saturated, and appears as a voltage generator. At the instant t_1, V_A reaches $+U$, the transistor saturates, and V_A remains fixed. For an additional time τ, however, V_B continues to increase, since the effect of the new conditions at A requires that long to propagate to B.

At the instant t_1, the wave v_r which has returned to A is totally reflected with change of sign, so that $v_i' = -v_r$. At B, from $(t_1 + \tau)$ to $(t_1 + 3\tau)$,

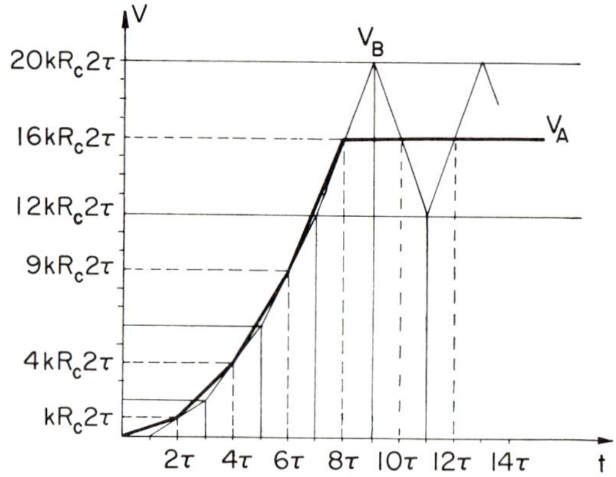

Fig. 27. Voltage waveforms in the circuit of Fig. 26.

v_i' (which is reflected without change of sign at B) is negative, and hence V_B decreases. At B from $(t_1 + 3\tau)$ to $(t_1 + 5\tau)$, the incident voltage wave v_i'', which is the wave reflected at A with change of sign, is positive (since the reflected wave v_r' is negative), and hence V_B increases. Thus there are oscillations at end B of the line (Fig. 27).

3.2 *Charging a line with* $R_0 = kR_c$, $R_l = \infty$

(a) STATEMENT OF THE PROBLEM (Fig. 28). Let us assume that the rise time of the transistor is $\ll \tau$ and that the line is lossless. At the instant 0^+

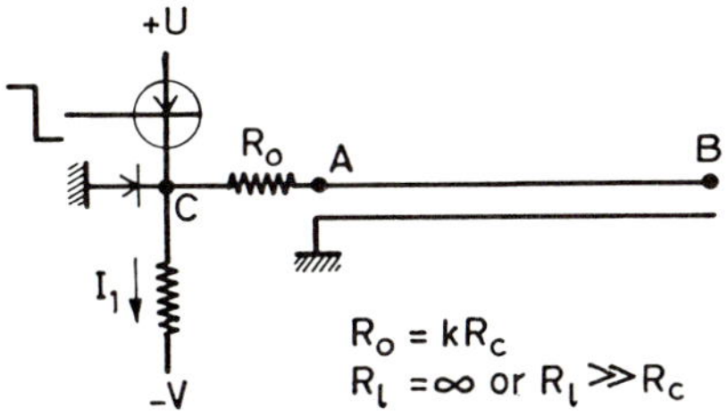

Fig. 28. Line charging with a nonideal source.

the transistor is conducting and $V_C = +U$ (saturated transistor). The reflection coefficient at A is

$$\Gamma_0 = \frac{R_0 - R_c}{R_0 + R_c}$$

(negative for $R_0 < R_c$), or, where $k = R_0/R_c$,

$$\Gamma_0 = \frac{k - 1}{k + 1}$$

(b) THE VOLTAGE (Fig. 29). At A, from 0 to 2τ, v_i is the voltage resulting from a voltage divider composed of R_0 and the line, which is equivalent to a resistance R_c:

$$V_A = 0 + v_i = U \cdot \frac{R_c}{R_0 + R_c} = U \frac{1}{k + 1}$$

or, since

$$\frac{1}{R_0 + R_c} = \frac{1 - \Gamma_0}{2R_c}, \qquad V_A = v_i = \frac{U}{2}(1 - \Gamma_0)$$

From τ to 3τ at B, v_i is reflected without change of sign, so that

$$v_r = v_i = +U \cdot \frac{R_c}{R_0 + R_c}; \quad V_B = v_i + v_r = \frac{2UR_c}{R_0 + R_c} = \frac{2U}{k + 1} = U(1 - \Gamma_0)$$

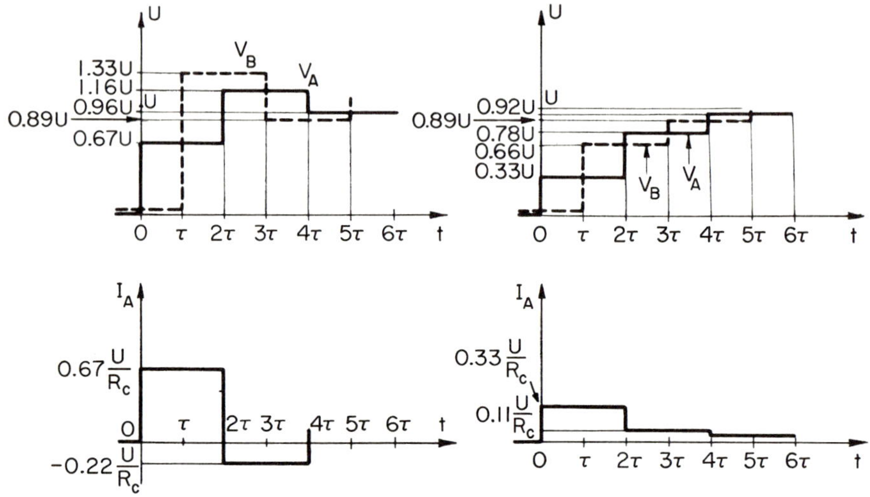

Case $R_0 < R_c$

$(R_0 = \frac{R_c}{2}, k = \frac{1}{2})$.

Case $R_0 > R_c$

$(R_0 = 2R_c, k = 2)$.

Fig. 29. Waveforms in the circuit of Fig. 28.

At A, from 2τ to 4τ, v_r is reflected and yields v_i',

$$v_i' = \Gamma_0 v_r$$

$$V_A = v_i + v_r + v_i' = v_i(1 + 1 + \Gamma_0) = \frac{U}{k+1}\left(2 - \frac{1-k}{1+k}\right) = U\frac{3k+1}{(k+1)^2}$$

or

$$V_A = \frac{U}{2}(2 - \Gamma_0 - \Gamma_0^2)$$

From 3τ to 5τ at B, the incident wave is again totally reflected without change of sign, and

$$V_B = v_i + v_r + v_i' + v_r' = 2v_i(1 + \Gamma_0) = U(1 - \Gamma_0^2)$$

At A, from 4τ to 6τ, since reflection occurs with coefficient Γ_0, superposition of the waves yields

$$V_A = v_i + v_r + v_i' + v_r' + v_i'' = v_i(1 + 1 + \Gamma_0 + \Gamma_0 + \Gamma_0^2)$$

$$V_A = \frac{U}{2}(2 - \Gamma_0^2 - \Gamma_0^3)$$

(c) THE CURRENT (Fig. 29). From 0 to 2τ,

$$I_A = i_i = \frac{v_i}{R_c} = \frac{U}{R_0 + R_c} = \frac{U(1 - \Gamma_0)}{2R_c}$$

from 2τ to 4τ,

$$I_A = i_i + i_r + i_i' = \frac{v_i}{R_c} - \frac{v_r}{R_c} + \frac{v_i'}{R_c}$$

$$I_A = \frac{v_i}{R_c}(1 - 1 + \Gamma_0) = \frac{U}{2R_c}(1 - \Gamma_0)\Gamma_0$$

and from 4τ to 6τ,

$$I_A = \frac{U}{2R_c}(1 - \Gamma_0)\Gamma_0^2$$

Thus V_A is not established instantaneously as in Section 3.1,1, but tends asymptotically towards U. If $R_0 > R_c$ ($k > 1$, $\Gamma_0 > 0$), V_A and V_B increase monotonically. If $R_0 < R_c$ ($k < 1$, $\Gamma_0 < 0$), V_A and V_B oscillate around U. The first overshoot of V_B (for $t = \tau$) is the smaller the nearer is R_0 to R_c.

An actual oscillogram of one of the situations depicted in Fig. 29 is shown in Fig. 29.1.

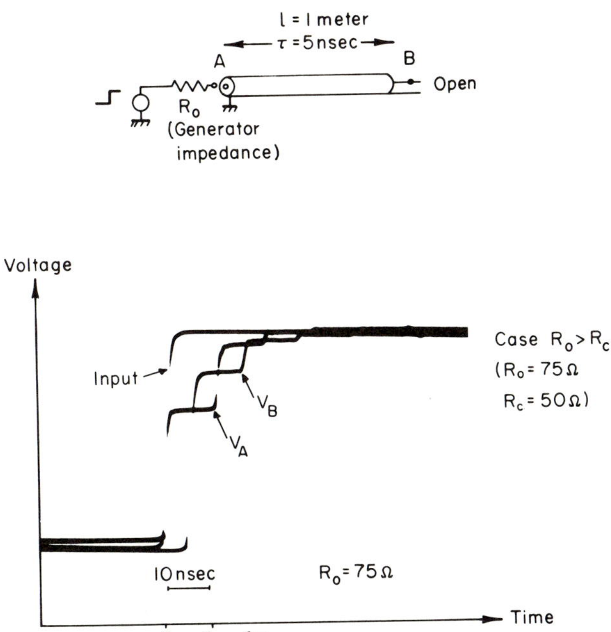

Fig. 29.1. Oscillogram corresponding to Fig. 29.

3.3 *Charging a line with* $R_0 = 0$, $R_l = kR_c$

(a) STATEMENT OF THE PROBLEM (Fig. 30). Let us assume that the rise time of the transistor is $\ll \tau$ and that the line is lossless. The reflection coefficient at B is

$$\Gamma_l = \frac{R_l - R_c}{R_l + R_c}$$

or, defining $k = R_l/R_c$,

$$\Gamma_l = \frac{k-1}{k+1}$$

At the instant 0^+ the transistor is conducting and $V_A = +U$. After this V_A remains constant (voltage generator).

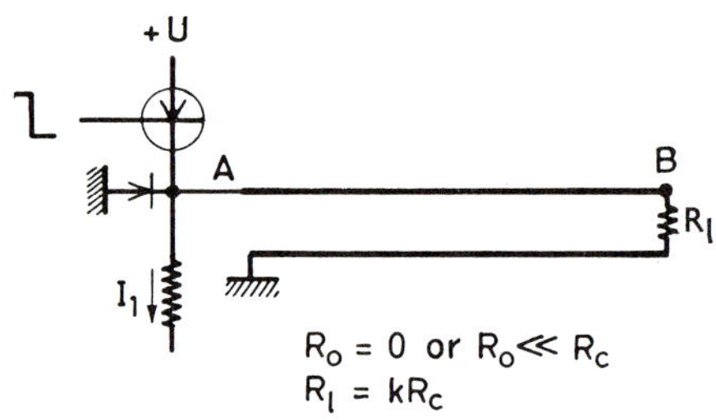

Fig. 30. Charging a loaded line.

(b) THE VOLTAGE (Fig. 31). At B, from τ to 3τ, $v_i = +U$ and $v_r = \Gamma_l v_i$, so that

$$V_B = v_i + v_r = U(1 + \Gamma_l)$$

or

$$V_B = U\frac{2k}{k+1}$$

At A, from 2τ to 4τ, v_r is totally reflected with change of sign (voltage generator), so that $v_i' = -v_r = -U\Gamma_l$. Thus as expected $V_A = +U$.

At B, from 3τ to 5τ,

$$V_B = v_i + v_r + v_i' + v_r' = U(1 + \Gamma_l - \Gamma_l - \Gamma_l^2)$$

$$V_B = U(1 - \Gamma_l^2) \quad \text{or} \quad V_B = U\frac{4k}{(k+1)^2}$$

At B from 5τ to 7τ, it will be found that

$$V_B = U(1 + \Gamma_l^3)$$

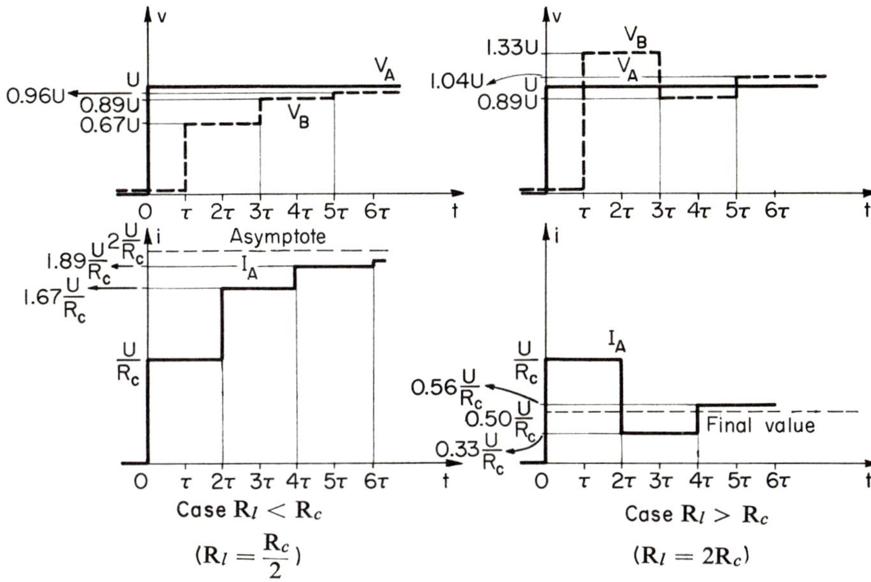

Fig. 31. Waveforms in Fig. 30.

(c) The Current (Fig. 31). From 0 to 2τ,

$$I_A = i_i = \frac{v_i}{R_c} = \frac{U}{R_c}$$

from 2τ to 4τ,

$$I_A = i_i + i_r + i_i' = \frac{v_i}{R_c} - \frac{v_r}{R_c} + \frac{v_i'}{R_c} = \frac{U}{R_c}(1 - 2\Gamma_l)$$

from 4τ to 6τ,

$$I_A = i_i + i_r + i_i' + i_r' + i_i'' = \frac{v_i}{R_c} - \frac{v_r}{R_c} + \frac{v_i'}{R_c} - \frac{v_r'}{R_c} + \frac{v_i''}{R_c}$$

$$I_A = \frac{U}{R_c}(1 - 2\Gamma_l + 2\Gamma_l^2)$$

The asymptotic value is U/R_l.

3.4 *Discharge of a line with* $R_0 = 0$, $R_l = \infty$

1. *The time required to cut off the transistor is* $\ll \tau$

(a) Statement of the Problem (Fig. 32). Assume the line is lossless. The line is discharged by a current generator, supplying a step of current I_1 in the negative direction with respect to the current in the line. At the

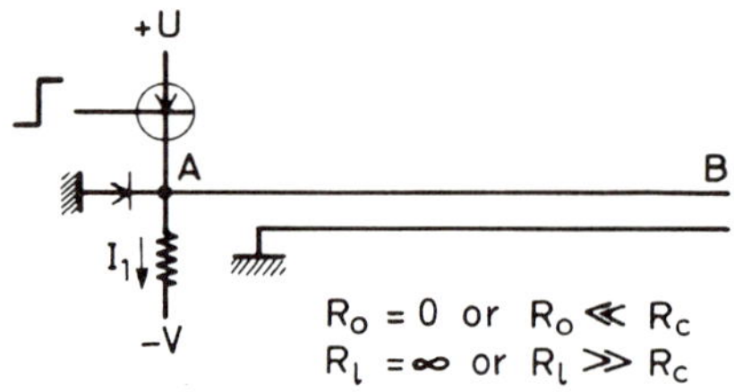

Fig. 32. Discharge of an open line.

beginning of the transient regime, the line appears as a resistance R_c. Two cases must be considered.

(b) $R_cI_1 < U$ (Fig. 33). At the initial time there exists an incident current wave

$$i_i = -I_1$$

so that

$$v_i = R_ci_i = -R_cI_1$$

Then at A, from 0 to 2τ, $V_A = U + v_i = U - R_cI_1$; at B, at the instant τ, v_i is reflected without change of sign ($R_l = \infty$), so that $v_r = v_i$; then at A,

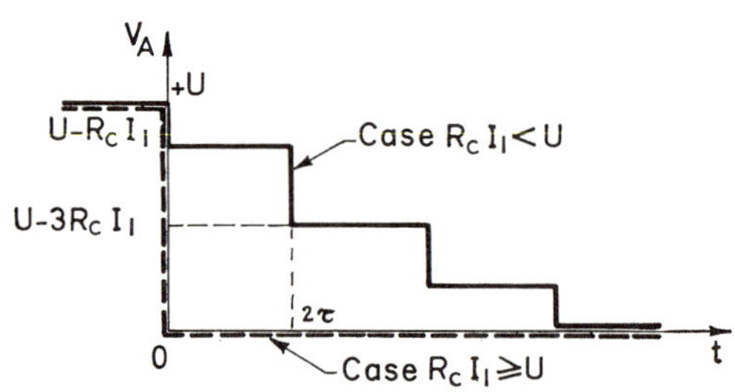

Fig. 33. Waveforms for circuit of Fig. 32.

from 2τ to 4τ, v_r is reflected from the current generator, which has an infinite internal impedance, even though $R_0 = 0$ (see the note at the beginning of Section 3), and thus

$$v_i' = v_r$$

$$V_A = U + v_i + v_r + v_i' = U - R_cI_1 - R_cI_1 - R_cI_1 = U - 3R_cI_1$$

Thus V_A decreases in steps until the voltage reaches zero, at which point the clamping diode operates. The voltage V_A then remains constant, and the clamping diode appears effectively as a voltage generator.

(c) $R_cI_1 \geq U$ (Fig. 33). The incident wave from the initial time is always equal to $-R_cI_1$, which would cause V_A to be negative from the beginning, except that the clamping diode operates and holds V_A at 0. The importance of having a sufficient current I_1 if the line is to be properly discharged can be seen.

2. *The collector current of the transistor decreases in a ramp, of duration* $> \tau$

(a) STATEMENT OF THE PROBLEM (Fig. 34). Beginning from the end of the transistor desaturation time, the collector current is

$$\frac{U}{R}(1 - kt)$$

The current through R flows both from the transistor and from the line (negative current in the line).

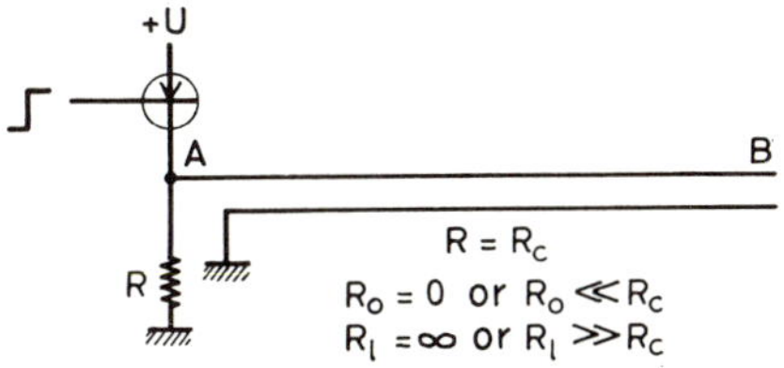

Fig. 34. Discharge of an open line.

(b) FIRST STAGE. The transistor is conducting in the normal region, and acts as a current generator (Fig. 35). From 0 to 2τ, the line appears as a resistance R_c with the extremity at a potential $+U$.

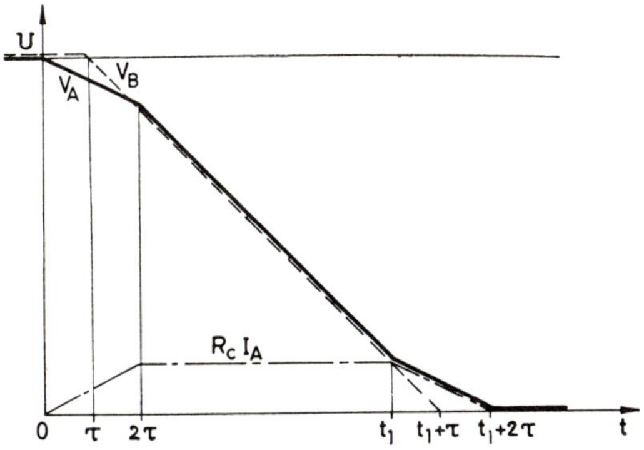

Fig. 35. Voltage waveforms for circuit of Fig. 34.

At A, from 0 to 2τ, we have, using the Kirchhoff current law,

$$\frac{V_A}{R} = \frac{U}{R}(1 - kt) + \frac{U - V_A}{R_c}$$

Thus if $R = R_c$, $V_A = U - (U/2)kt$, and V_A decreases with slope $kU/2$. Further, since $V_A = U + v_i$, $v_i = -Ukt/2$.

At B, starting at time τ,

$$v_r = v_i = -\frac{U}{2}k(t - \tau)$$

so that $V_B = U + v_i + v_r = U[1 - k(t - \tau)]$. Thus V_B decreases with slope kU.

At A, starting at time 2τ, the wave v_r sees the parallel combination of $R = R_c$ and a current generator (very high resistance). Thus the line is matched at the input (even though $R_0 = 0$, see the note at the beginning of Section 3). There is thus no reflection, and since

$$V_A = U + v_i + v_r; \qquad v_i = -\frac{U}{2}kt$$

$$v_r = -\frac{U}{2}k(t - 2\tau)$$

(the shift in v_r is due to the transmission time), we have

$$V_A = U[1 - k(t - \tau)]$$

Thus V_A decreases with slope kU.

(c) Second Stage. The collector current reaches zero, and the transistor cuts off, at an instant t_1 (Fig. 35), given by $U(1 - kt_1)/R = 0$, from which $t_1 = 1/k$. However V_A is not zero at t_1 ($V_A = Uk\tau$).

Starting from time t_1, the current law applied at A shows that the currents in R and in the line are equal (that in the line is negative): $V_A/R = -I_A$. With $R = R_c$, this yields

$$\frac{U + v_i + v_r}{R_c} = -i_i - i_r = -\frac{v_i}{R_c} + \frac{v_r}{R_c}$$

from which $v_i = -U/2$. Thus the incident wave is constant.

For $t_1 < t < t_1 + 2\tau$, v_r is the same as in the first stage, so that

$$V_A = U + v_i + v_r = U - \frac{U}{2} - \frac{U}{2}k(t - 2\tau) = \frac{U}{2}[1 - k(t - 2\tau)]$$

Thus V_A decreases with slope $kU/2$, and reaches zero at the time $(t_1 + 2\tau)$.

At B, there is no change from the situation of the first stage up till time $t_1 + \tau$. After that, $v_i = \text{const} = -U/2$, and $v_r = v_i$, so that $V_B = U + v_i + v_r = 0$, and the line is discharged.

3.5 *Line matched at the input and terminated in an arbitrary impedance*

Even though in defining the reflection coefficients at the beginning of this chapter, we assumed the lines would be terminated in pure resistances, we can treat some simple cases with complex impedances, anticipating some results from Chapter IV.

(a) Statement of the Problem (Fig. 36). Suppose that the time required to cut off the transistor is $\ll \tau$, and that the line is terminated in a

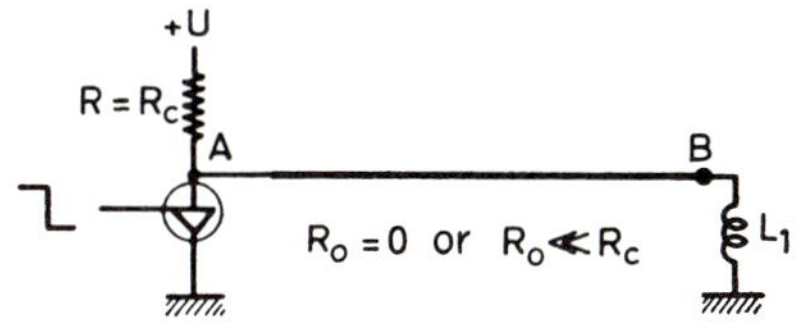

Fig. 36. Line with inductive load.

pure self-inductance. Starting with the initial instant, the current in R is equal to that in the line.

(b) Evolution of the Voltage (Fig. 37). At A, from 0 to 2τ, the line appears as a resistance R_c. The current law yields $(U - V_A)/R = V_A/R_c$. Since $R = R_c$, this results in $v_i = V_A = U/2$.

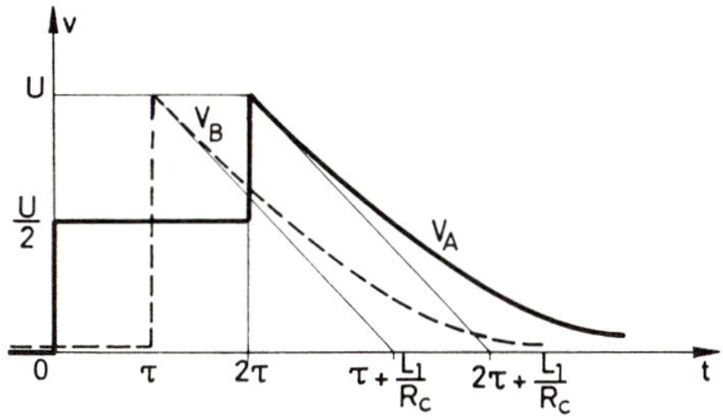

Fig. 37. Voltage waveforms in circuit of Fig. 36.

At B, beginning at time τ,

$$V_B = L_1 \frac{dI_B}{dt}; \qquad v_i + v_r = L_1\left(\frac{di_i}{dt} + \frac{di_r}{dt}\right) = \frac{L_1}{R_c}\left(\frac{dv_i}{dt} - \frac{dv_r}{dt}\right)$$

Since $v_i = \mathrm{U}/2 = \text{const}$, $dv_i/dt = 0$, and thus

$$\frac{dv_r}{\mathrm{U}/2 + v_r} = -\frac{\mathrm{R}_c}{\mathrm{L}_1}\,dt; \qquad v_r = -\frac{\mathrm{U}}{2} + \mathrm{K}\exp[-t/(\mathrm{L}_1/\mathrm{R}_c)]$$

where K is some constant. At time $t = \tau$, L_1 appears as an infinite imped-

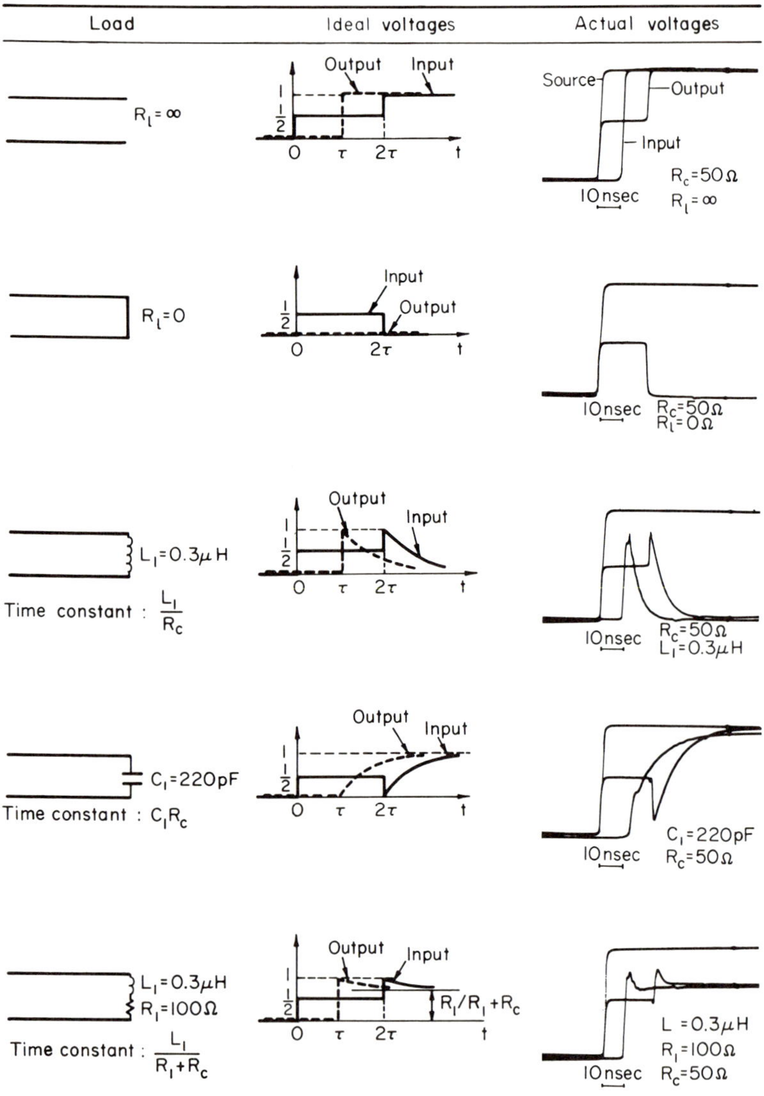

Fig. 38. Voltage step response of various lines. Length 2 meters, lossless, characteristic resistance 50 Ω, matched source.

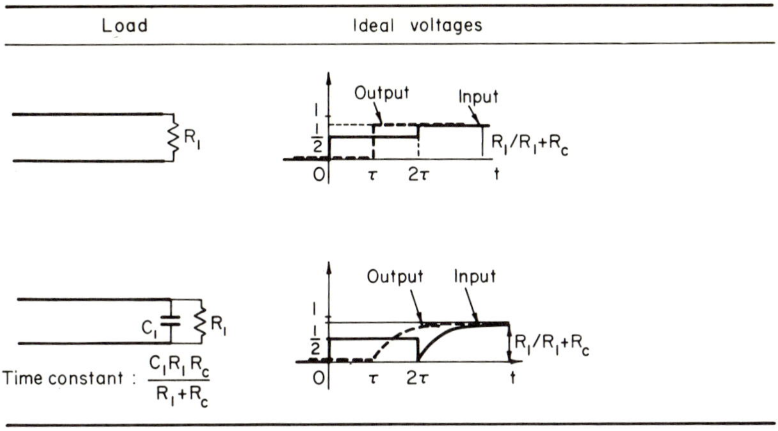

Fig. 38 *(continued)*

ance, so that $v_i(\tau) = v_r(\tau)$, from which

$$K = U \exp[\tau/(L_1/R_c)]$$

$$v_r = -\frac{U}{2} + U \exp[-(t - \tau)/(L_1/R_c)]$$

$$V_B = v_i + v_r = U \exp[-(t - \tau)/(L_1/R_c)]$$

At A, beginning at 2τ, v_r sees a matched impedance (the transistor internal resistance is very large; see the remark at the beginning of Section 3), and thus is not reflected. Thus $V_A = v_i + v_r$, where v_r is the value taken at B, but shifted in time, so that

$$V_A = U \exp[-(t - 2\tau)/(L_1/R_c)]$$

(c) Other Examples. Figure 38 shows the responses of lines, each of which is lossless, of length l and characteristic impedance R_c, fed by a source with internal impedance R_c supplying a unit step of voltage, but each of which is terminated in a different impedance. (R_1 can be a lumped resistance or the characteristic impedance of another transmission line.) Recall that $R_c = (L/C)^{1/2}$, $\tau = l(LC)^{1/2}$, where L and C are the per unit length self-inductance and capacitance of the line.

3.6 *Line terminated in a distributed impedance. Reflection and transmission coefficient*

The relations governing the transition from one transmission line to another can be obtained by writing the Kirchhoff laws at the junction (Fig. 40).

(a) Sign Conventions. It is essential to distinguish between the direction of current flow and the direction of propagation of the current wave (Fig. 39). We will use the convention that the arrow indicates the direction

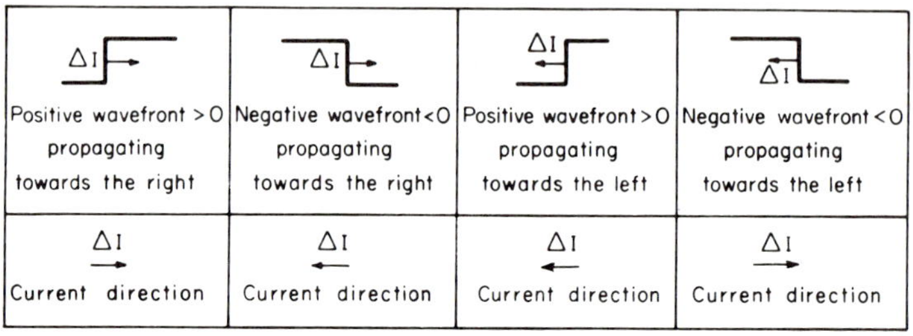

Fig. 39. Direction of conventional current for various waves.

of propagation (not to be confused with the direction of current flow) in the forward conductor.

We will now consider two lines with characteristic impedances R_{c1} and R_{c2}. The positive senses for the various currents and voltages are shown in Fig. 40.

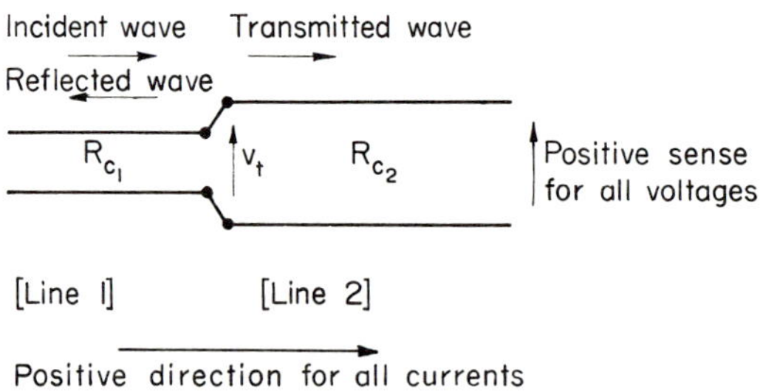

Fig. 40. Line with discontinuous structure.

(b) Fundamental Relations. Suppose that an incident step (v_i, i_i) propagates from left to right. When it arrives at the discontinuity of characteristic impedance, it gives rise to a reflected step (v_r, i_r) and a transmitted step (v_t, i_t).

Three relations result from the fundamental property of transmission lines:

$$v_i/i_i = R_{c1} \tag{II.4}$$

$$v_t/i_t = R_{c2} \tag{II.5}$$

$$v_r/i_r = -R_{c1} \tag{II.6}$$

The waves v_i and i_i have the same sign, as do v_t and i_t, but v_r and i_r have the opposite sign.

At the junction of the two lines,

$$i_t = i_i + i_r \tag{II.7}$$

$$v_t = v_i + v_r \tag{II.8}$$

Relation (II.7) results from the Kirchhoff current law, with the conventions adopted for the positive current direction. Relation (II.8) states the fact that at the junction the voltage across line 1, $v_i + v_r$, is the same as that across line 2, v_t.[†]

Note: relations (II.7) and (II.8) express the fact that energy is conserved at the junction of the two lines:

$$|\, v_i i_i \,| = |\, v_r i_r \,| + |\, v_t i_t \,| \tag{II.9}$$

(Energy being always positive, it is necessary to take the absolute values of the various terms.)

Relation (II.9) can be written

$$v_i i_i = v_t i_t + |\, v_r i_r \,|$$

which yields, taking account of relations (II.4)–(II.7):

$$\frac{v_i^2}{\mathrm{R}_{c1}} = v_t(i_i + i_r) + \frac{v_r^2}{\mathrm{R}_{c1}}$$

$$= v_t\left(\frac{v_i}{\mathrm{R}_{c1}} - \frac{v_r}{\mathrm{R}_{c1}}\right) + \frac{v_r^2}{\mathrm{R}_{c1}}$$

from which

$$v_i^2 - v_r^2 = v_t(v_i - v_r); \qquad v_i + v_r = v_t$$

(c) THE REFLECTION AND TRANSMISSION COEFFICIENTS. Let us introduce a reflection coefficient Γ defined by

$$\Gamma = \frac{\mathrm{R}_{c2} - \mathrm{R}_{c1}}{\mathrm{R}_{c2} + \mathrm{R}_{c1}} \tag{II.10}$$

Γ is > 0 if $\mathrm{R}_{c2} > \mathrm{R}_{c1}$, and Γ is < 0 if $\mathrm{R}_{c2} < \mathrm{R}_{c1}$. Writing v_r and v_t as functions of v_i, R_{c1} and R_{c2}:

$$v_r = v_t - v_i = \mathrm{R}_{c2} i_t - v_i = \mathrm{R}_{c2}(i_i + i_r) - v_i$$

$$v_r = \mathrm{R}_{c2}\left(\frac{v_i}{\mathrm{R}_{c1}} - \frac{v_r}{\mathrm{R}_{c1}}\right) - v_i$$

[†] It is well to consider in (v_i, i_i) not only the wave emanating directly from the generator at the input to line 1, but also the waves reflected at the input (after having been reflected at the junction and at the load). The same remark applies to (v_r, i_r).

These yield

$$v_r\left(1 + \frac{R_{c2}}{R_{c1}}\right) = v_i\left(\frac{R_{c2}}{R_{c1}} - 1\right)$$

$$v_r = v_i \cdot \frac{R_{c2} - R_{c1}}{R_{c2} + R_{c1}}$$

$$v_r = \Gamma v_i \tag{II.11}$$

Thus Γ is the voltage reflection coefficient from line 1 to line 2.

In addition,

$$v_t = v_i + v_r = v_i(1 + \Gamma)$$

$$v_t = (1 + \Gamma)v_i \tag{II.12}$$

Thus $(1 + \Gamma)$ is the voltage transmission coefficient from line 1 to line 2.

The voltage reflection coefficient from line 2 to line 1 is

$$-\Gamma = \frac{R_{c1} - R_{c2}}{R_{c1} + R_{c2}}$$

The voltage transmission coefficient from line 2 to line 1 is $(1 - \Gamma)$.

A simple calculation for the currents shows that

$$i_r = -\Gamma i_i \tag{II.13}$$

$$i_t = (1 - \Gamma)i_i \tag{II.14}$$

That is, for the currents, the reflection coefficient is $-\Gamma$ and the transmission coefficient is $1 - \Gamma$, with the sign conventions indicated in Section 3.6 (a).

(d) Signal of Arbitrary Form (Fig. 41). All the results established above for step functions remain valid for waves of any form whatever,

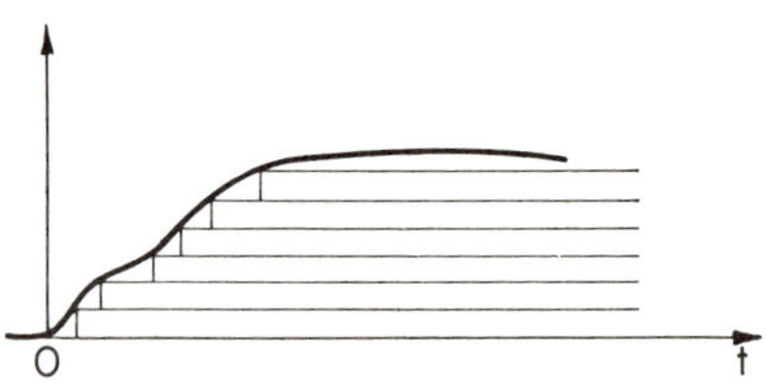

Fig. 41. Representation of an arbitrary waveform as a sum of steps.

since an arbitrary waveform can be approximated by a superposition of step functions.

3.7 *Matching by means of parallel matching stubs* (Fig. 42)

It is not always possible to match a load to a line using lumped impedances. In such a case, a procedure well known in waveguide and uhf transmission can be used.

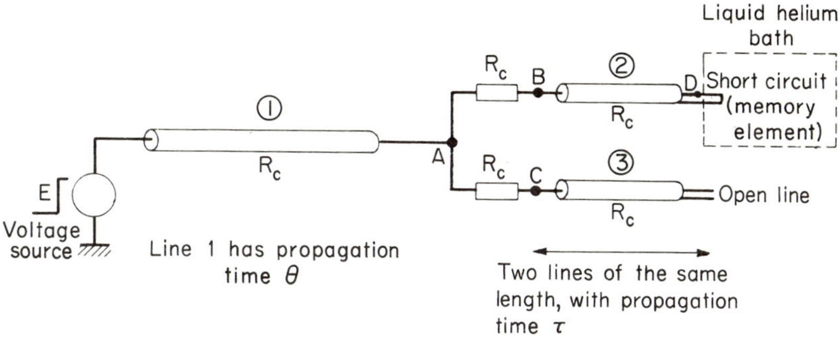

Fig. 42. Matching a line using a parallel stub. Line 1 can be of any length, including zero.

Consider a line terminated in a short circuit, such as a cryogenic memory element. It is not possible to place a resistance R_c at the end of the line, because in this case the end is immersed in a bath of liquid helium, and, in order to limit evaporation of the helium, it is desirable to cause as little energy dissipation in the bath as possible. Yet it is still desirable to prevent the reflection produced at the short circuit from returning to the generator.

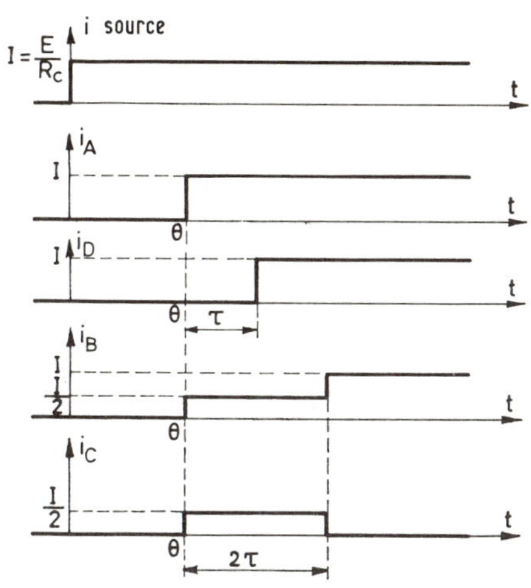

Fig. 43. Current waveforms in circuit of Fig. 42.

One possible solution to this problem is shown in Fig. 42. The wave front coming from the source encounters at A an impedance R_c. Thus there is no reflection in line 1 towards the source. The incident wave at A divides into two waves of the same amplitude which propagate into lines 2 and 3. These two waves are reflected at the ends of lines 2 and 3 and return to A. After this two-way trip through lines 2 and 3, the voltage wave reflected at the short circuit has changed sign, while the voltage wave

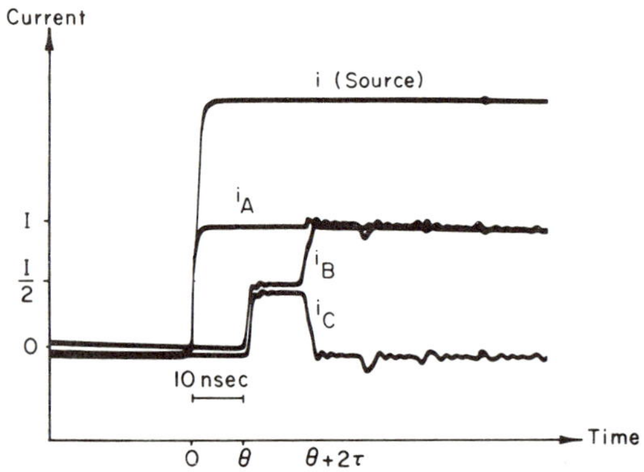

Fig. 43.1. Oscillograms of waveforms of Fig. 43.

reflected at the open circuit returns with the same sign. These two reflected waves thus exactly cancel at A, and nothing is transmitted back to the source. Because of the matched conditions at B and C, neither returning wave is reflected back towards the outputs of lines 2 and 3.

The final current $I = E/R_c$ is established in the short circuit immediately upon arrival of the first wave front. The source furnishes the current I immediately (Figs. 43 and 43.1).

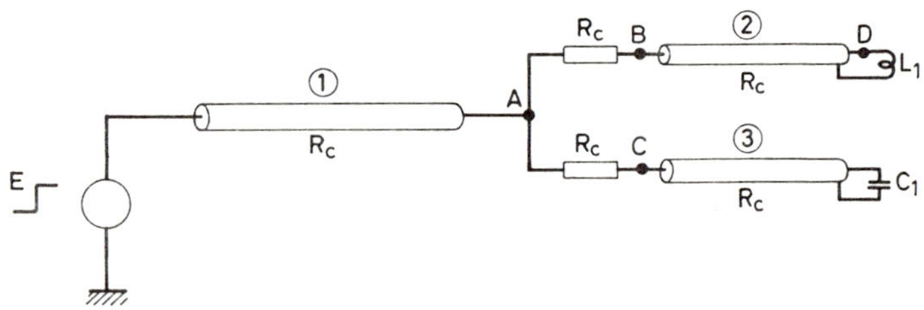

Fig. 44. Matching a reactive load using a stub.

In the same way, it is also possible to compensate for a self-inductance at the end of line 2, by a capacitance at the end of line 3, and vice versa (Fig. 44) (see Exercise 3 at the end of this chapter).

4. A Tabular Method

The tabular method to be presented below is a numerical presentation of the method of traveling waves. It allows study of the time variation of the voltage at each point of a transmission line, or of a combination of transmission lines with reflection coefficients different from zero. The method amounts to arranging the significant results of the analysis in a table, according to certain rules which systematize and order the study.

4.1 *Explanation of the method*

The method will be explained using a simple example (Fig. 45). Suppose that a line with characteristic resistance $R_c = 50\,\Omega$, terminated in a resistance $R_l = 100\,\Omega$, is excited at some initial instant by a unit step of

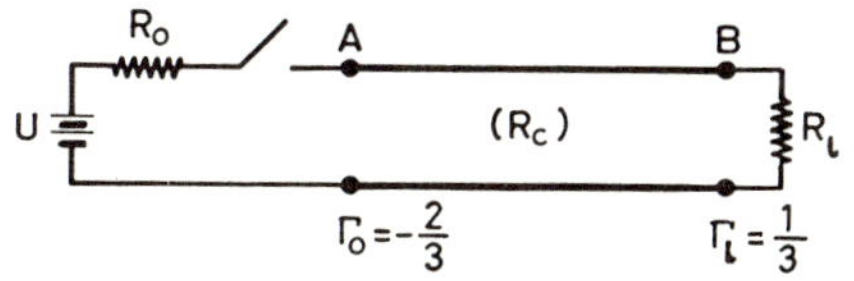

Fig. 45. Circuit used to illustrate the tabular method.

voltage arising in a generator with internal resistance $R_0 = 10\,\Omega$. Thus the internal voltage source of the generator is

$$U = 1 \cdot \frac{50 + 10}{50} = 1.2\ \text{V}$$

The reflection coefficients are

$$\Gamma_l = \frac{R_l - R_c}{R_l + R_c} = \frac{1}{3}; \qquad \Gamma_0 = \frac{R_0 - R_c}{R_0 + R_c} = -\frac{2}{3}$$

Let us set up a table having three columns (Fig. 46), and place on the first line, corresponding to time 0, the state of the transmission line at the instant it is first excited at its input by a *unit* step of voltage. Thus in the "input" column we place a 1, and in the "output" column a 0.

At time τ, the step will have arrived at the output, and we write on the second line, corresponding to time τ, the number 1 in the output column.

At this instant (τ), and at this place (the load), the incident wave is reflected with $\Gamma_l = \frac{1}{3}$. Thus a step of amplitude $\frac{1}{3}$ is reflected back towards the input, and the voltage at the output is modified, at time τ, by $\frac{1}{3}$, which number we write below the 1 written previously in the output column.

Thus the voltage at the output of the line, starting at time τ, is $1 + \frac{1}{3} = \frac{4}{3}$, which is written as indicated in the table.

In order to show better that a wave is transmitted from the input of the line towards the output between the times 0 and τ, an inclined arrow is drawn to indicate the transmission path of the wave (this arrow will be of importance later). In addition, a circle is drawn around the voltages at the input and output of the line each time the voltage at one of these points changes. Thus the 1 in the input column and the $\frac{4}{3}$ in the output column are circled in the table.

Let us consider now the reflected wave, of amplitude $\frac{1}{3}$, which travels towards the input. At time 2τ, this wave arrives at the input with amplitude $\frac{1}{3}$, and is reflected with reflection coefficient $\Gamma_0 = -\frac{2}{3}$. The reflected wave thus has amplitude $\frac{1}{3} \times -\frac{2}{3} = -\frac{2}{9}$. This wave travels back towards the output, where it arrives at time 3τ. The change in amplitude at the input to the line is $\frac{1}{3} + -\frac{2}{9} = \frac{1}{9}$. These quantities are written in the table as

Time	Input		Output
0	(1)	→	
τ			1, $\frac{1}{3}$ } $\left(\frac{4}{3}\right)$
		←	
2τ	$\left(\frac{1}{9}\right)$ { $\frac{1}{3}$, $-\frac{2}{9}$		
		→	
3τ			$-\frac{2}{9}$, $-\frac{2}{27}$ } $\left(-\frac{8}{27}\right)$
		←	
4τ	$\left(-\frac{2}{81}\right)$ { $-\frac{2}{27}$, $\frac{4}{81}$		
		→	
5τ			$\frac{4}{81}$, $\frac{4}{243}$ } $\left(\frac{16}{243}\right)$
		←	
6τ	$\left(\frac{4}{279}\right)$ { $\frac{4}{243}$, $-\frac{8}{729}$		
		→	
7τ			$-\frac{8}{729}$, $-\frac{8}{2187}$ } $\left(-\frac{32}{2187}\right)$
		←	

Fig. 46. Table corresponding to the circuit of Fig. 45.

indicated, the $\frac{1}{9}$ is circled, and an arrow descending towards the left is drawn between the lines τ and 2τ. Note that starting from time 2τ, the amplitude at the input is $1 + \frac{1}{9}$. The quantity $\frac{1}{9}$ placed in the table is only the change in amplitude.

The same procedure is followed for the other lines of the table, to produce the results shown in Fig. 46.

Some checks can be made of the numbers in the table. For a lossless line, at $t = \infty$ the input and output voltages must be equal. But it is easy to see from the circuit that the line voltage for $t = \infty$ is $1.2 \times (100/110) = 1.09$ volts. Thus we should have at the input

$$1.09 = 1 + \frac{1}{9} - \frac{2}{81} + \frac{4}{729} - \frac{8}{6561} + \cdots + (-1)^n \frac{2^n}{9^{n+1}} + \cdots$$

and at the output

$$1.09 = \frac{4}{3} - \frac{8}{27} + \frac{16}{243} - \frac{32}{2187} + \frac{64}{19{,}683} - \cdots + (-1)^n \frac{4 \cdot 2^n}{3^{2n+1}} + \cdots$$

These can be verified, using the formula for the sum of a geometric series with ratio $-\frac{2}{9}$.

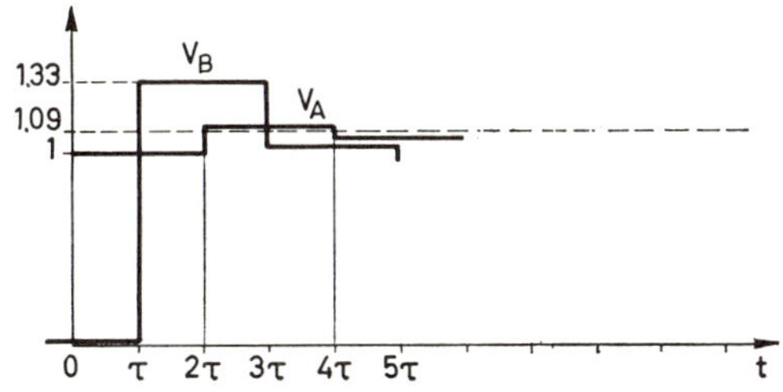

Fig. 47. Voltage waveforms in the circuit of Fig. 45.

The various signals obtained by this method are shown in Fig. 47. The precision obtainable with this tabular method can be as great as desired.

4.2 *Generalization* (Fig. 48)

The preceding table can be generalized by writing in the literal formulas for the various quantities, in which case the table of Fig. 48 results.

At $t = \infty$, it is possible to show that the two series obtained by summing the terms in the boxes in Fig. 48 converge to the same limit:

$$\frac{1 + \Gamma_l}{1 - \Gamma_0 \Gamma_l}$$

Time	Input			Output
0	$\boxed{1}$		→	
τ			1 Γ_l	$\boxed{1+\Gamma_l}$
2τ	$\boxed{\Gamma_l(1+\Gamma_0)}$	Γ_l $\Gamma_l\Gamma_0$	← →	
3τ			$\Gamma_l\Gamma_0$ $\Gamma_l^2\Gamma_0$	$\boxed{\Gamma_l\Gamma_0(1+\Gamma_l)}$
4τ	$\boxed{\Gamma_l^2\Gamma_0(1+\Gamma_0)}$	$\Gamma_l^2\Gamma_0$ $\Gamma_l^2\Gamma_0^2$	← →	
5τ			$\Gamma_l^2\Gamma_0^2$ $\Gamma_l^3\Gamma_0^2$	$\boxed{\Gamma_l^2\Gamma_0^2(1+\Gamma_l)}$
⋮			←	
∞	Sum of the above terms $1+\Gamma_l(1+\Gamma_0)+\Gamma_l^2\Gamma_0(1+\Gamma_0)+\cdots$			Sum of the above terms $1+\Gamma_l+\Gamma_l\Gamma_0(1+\Gamma_l)$ $+\Gamma_l^2\Gamma_0^2(1+\Gamma_l)+\cdots$

Fig. 48. Literal form of the table of Fig. 46.

If in this Γ_l and Γ_0 are replaced by their values given above, there results

$$\frac{1 + (R_l - R_c)/(R_l + R_c)}{1 - \{(R_l - R_c)/(R_l + R_c)\}\{(R_0 - R_c)/(R_0 + R_c)\}}$$

This is none other than the final voltage on the line, as it should be, if it is noted that the battery voltage which yields 1 V at the input to the line at the initial time is

$$\frac{R_c + R_0}{R_c}.$$

The main interest of this tabular method, aside from the precision obtainable, is that it applies also to extremely complex cases of lines in combination, or lines with numerous discontinuities.

4.3 *Examples*

(a) First Example. Consider the line shown in Fig. 49.

The battery voltage needed to produce a step of 1 volt at the input to the line is $(10 + 50)/50 = 1.2$ volts.

To find the reflection coefficient at the discontinuity in the middle of the line, note that the incident wave arriving at M sees in parallel the 100-Ω resistance and the 50-Ω characteristic resistance of the portion of the line

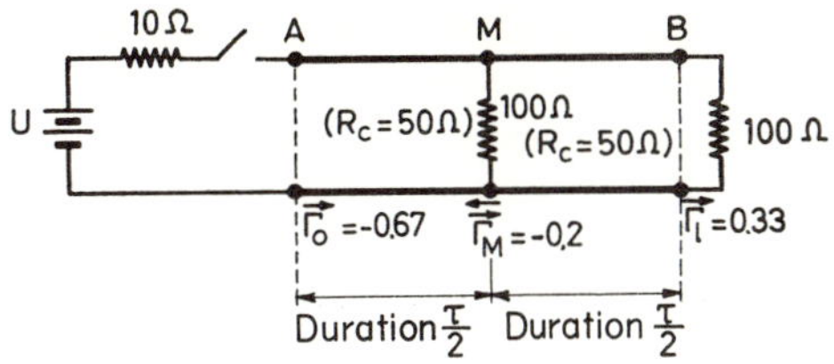

Fig. 49. Line having a lumped element at its midpoint.

from M to B. This is an equivalent resistance of 33.33 Ω, and the reflection coefficients are then

$$\Gamma_0 = \frac{10-50}{10+50} = -0.67; \quad \Gamma_{\mathrm{M}} = \frac{33.33-50}{33.33+50} = -0.2; \quad \Gamma_l = \frac{100-50}{100+50} = 0.33$$

A table with four columns is constructed (Fig. 50).

The unit step arrives at the middle of the line at time $\tau/2$, and is reflected. The reflected wave has amplitude -0.2, and the transmitted wave, which

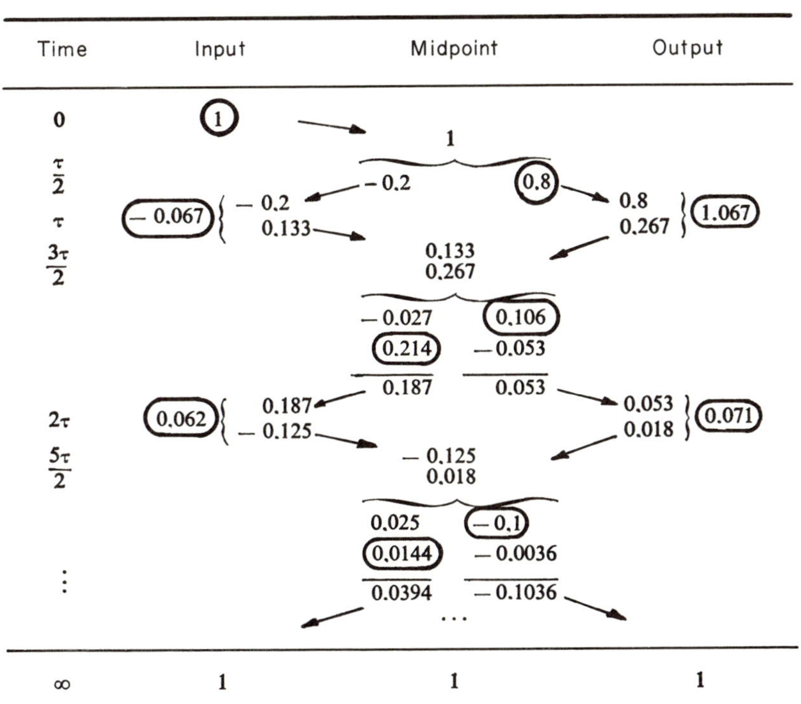

Fig. 50. Table corresponding to the circuit of Fig. 49.

is the sum of the incident and reflected waves, has amplitude 0.8. Each of these waves then propagates, the former towards the left, and the latter towards the right. Corresponding arrows are placed in the table, and the number 0.8, representing the change in voltage at the middle of the line, is circled.

At time τ, the waves -0.2 and 0.8 are reflected, and the reflected waves have amplitudes, respectively, 0.133 and 0.267. The changes in the input and output voltages are, respectively, -0.067 and 1.067.

These new reflected waves then both propagate towards the middle of the line, where they arrive together at time $3\tau/2$. Thus in the "middle" column of the table, the numbers 0.133 and 0.267 are written one under the other. These numbers can not, however, simply be added, because the corresponding waves are not propagating in the same direction. It is necessary to treat each wave separately. Then there are obtained successively the numbers -0.027 and 0.106 for the one wave, and -0.053 and 0.214 for the other wave, the two reflected waves being -0.027 and -0.053. It is now possible to add the waves propagating in the same direction, with the result that the wave propagating towards the left has amplitude 0.187, while that propagating towards the right has amplitude 0.053. The numbers 0.106 and 0.214, which are the transmitted waves, are circled in the table.

The rest of the table is filled in the same way.

It is easy to check that the three series thus obtained all sum to unity, which is the voltage on the line for $t = \infty$, which is found from the battery voltage (1.2) and the voltage divider formed by the 10-Ω and 100-Ω resistances.

(b) Second Example (Fig. 51). Consider two lines with different characteristic resistances, placed end to end.

The method is the same as in the preceding example, except that in the middle of the line the reflection coefficients have opposite signs for the two directions of transmission (Fig. 51). Thus, where the arrow indicates the direction of propagation

$$\Gamma_0 = -0.5; \qquad \overrightarrow{\Gamma} = -0.2; \qquad \overleftarrow{\Gamma} = 0.2; \qquad \Gamma_l = -0.667$$

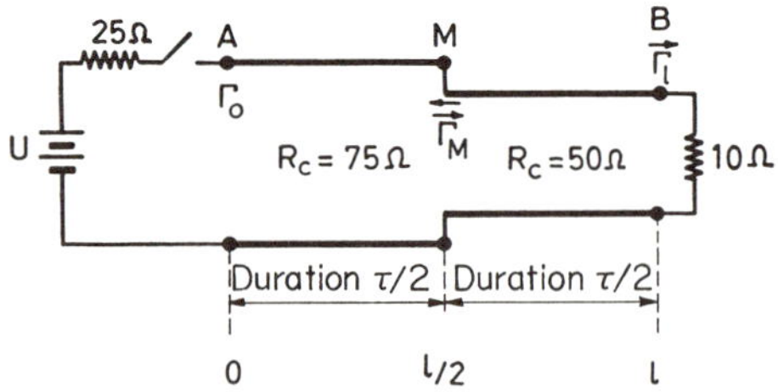

Fig. 51. Line with discontinuous characteristic resistance.

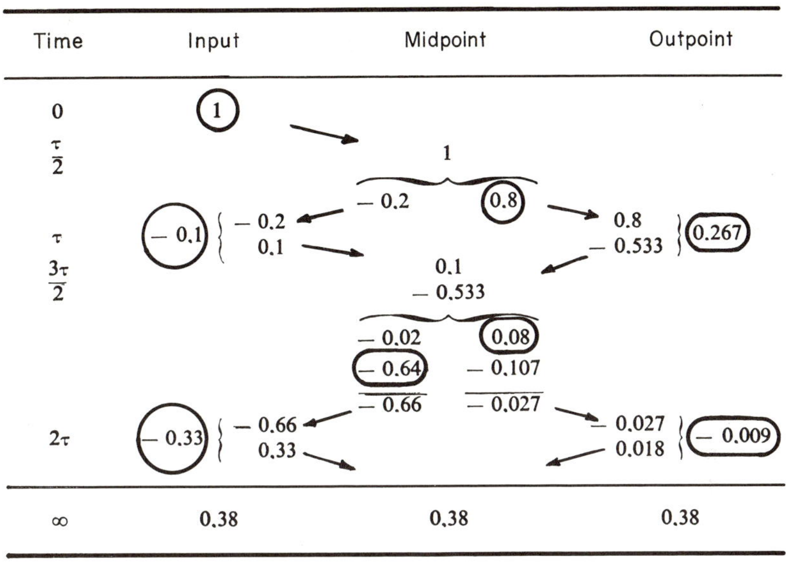

Fig. 52. Table corresponding to the circuit of Fig. 51.

The battery voltage which results in a voltage at A of 1 volt at the initial instant is U = 100/75 = 1.333 volts. The line voltage for $t = \infty$ is

$$\frac{1.333 \times 10}{10 + 25} = 0.38 \text{ volts}$$

The resulting table is that shown in Fig. 52.

Remark. As is apparent, quite varied cases can be treated using this tabular method. The same table can also be constructed starting with a step of current rather than with a voltage step. Only the signs of the reflection coefficients are changed in that case. Finally, the variation of the voltage at any arbitrary point along the line can also be found by this method.

5. Appendix. Formulas for the Method of Traveling Waves in the General Case of a Line with Losses

5.1 *Introduction*

In the development to follow, the equations for lossy lines will be treated using the Laplace transformation. In this, the results and formulas established in Chapter I will be used as necessary.

It should be mentioned that, according to a result to be established in Chapter IV, the propagation velocity of a wave depends on time, since the propagation function for a line with loss is not a constant.

The relations $v = v_i + v_r$, $i = i_i + i_r$, which we will establish, will apply when the currents and voltages on a transmission line can be considered as the superposition of a double infinity of incident and reflected traveling waves. These waves propagate in opposite directions. In addition, the propagation velocities of the waves are slightly different from one wave to another, except in the cases of a line without dispersion (in which case the Heaviside relation is satisfied), and a line without loss (the case considered previously).

In the cases of lines without dispersion and lines without loss, all the incident waves can be superposed, and then all the reflected waves. The result is two voltage waves propagating in opposite directions with the same velocity

$$u = \frac{1}{(\mathrm{LC})^{1/2}} = \frac{1}{\delta}$$

With each voltage wave there is associated a current wave, the ratio of the amplitudes of the two waves being the characteristic resistance of the line $\mathrm{R}_c = (\mathrm{L/C})^{1/2}$.

The form of the general equations involved makes this decomposition necessary in the sinusoidal steady state, as well as in the transient regime, regardless of the shape of the waves. We will thus derive a method independent of the form of the excitation applied.

5.2 *Summary of the relations used*

Consider a line initially at rest, to which at the initial time $t = 0$ an arbitrary excitation is applied. In Chapter I it was seen that the subsequent operation is governed by the relations

$$\mathrm{V}(x, p) = \mathrm{V}_1 e^{-\gamma(p)x} + \mathrm{V}_2 e^{+\gamma(p)x}$$

$$\mathrm{I}(x, p) = \frac{\mathrm{V}_1}{\mathrm{Z}_c} e^{-\gamma(p)x} - \frac{\mathrm{V}_2}{\mathrm{Z}_c} e^{+\gamma(p)x}$$

where

$$\mathrm{Z}_c = \left(\frac{\mathrm{R} + \mathrm{L}p}{\mathrm{G} + \mathrm{C}p}\right)^{1/2}; \qquad \mathrm{V}_1 = \mathrm{E}(p)\cdot\frac{\mathrm{Z}_c(p)}{\mathrm{Z}_0(p) + \mathrm{Z}_c(p)}\cdot\frac{1}{1 - \Gamma_0\Gamma_l e^{-2\gamma l}}$$

$$\gamma = [(\mathrm{R} + \mathrm{L}p)(\mathrm{G} + \mathrm{C}p)]^{1/2}; \quad \mathrm{V}_2 = \mathrm{E}(p)\cdot\frac{\mathrm{Z}_c(p)}{\mathrm{Z}_0(p) + \mathrm{Z}_c(p)}\cdot\frac{\Gamma_l e^{-2\gamma l}}{1 - \Gamma_0\Gamma_l e^{-2\gamma l}}$$

Let us write

$$V_1(p) = M(p); \qquad M(p) = E(p) \cdot \frac{Z_c(p)}{Z_0(p) + Z_c(p)} \cdot \frac{1}{1 - \Gamma_0 \Gamma_l e^{-2\gamma l}}$$

$$V_2(p) = N(p) e^{-2\gamma l}; \qquad N(p) = E(p) \cdot \frac{Z_c(p)}{Z_0(p) + Z_c(p)} \cdot \frac{\Gamma_l}{1 - \Gamma_0 \Gamma_l e^{-2\gamma l}}$$

where $M(p)$ and $N(p)$ are the Laplace transforms of certain functions $m(t)$ and $n(t)$.

Three cases of interest will be examined: (1) the line with only small losses at the beginning of the transient regime; (2) the line without dispersion (Heaviside condition satisfied) (this case is theoretical, since such lines do not exist, except in special circumstances); (3) the lossless line, which is a special case of the preceding case.

These three models can serve as a first approximation to real lines.

(a) Lines with Small Losses at the Beginning of the Transient Regime. If t_1 is the smaller of L/R and C/G, the beginning of the transient regime is $0 < t \ll t_1$. By a line with small losses is meant a line for which $R/(Lp) \ll 1$ and $G/(Cp) \ll 1$. In either case,

$$\begin{aligned}
\gamma &= [(R + Lp)(G + Cp)]^{1/2} = (LCp^2)^{1/2}[(1 + R/Lp)(1 + G/Cp)]^{1/2} \\
&= p(LC)^{1/2}(1 + R/Lp + G/Cp + RG/LCp^2)^{1/2} \\
&\approx p(LC)^{1/2}[1 + \tfrac{1}{2}(R/Lp + G/Cp)] \\
&= p(LC)^{1/2} + \tfrac{1}{2}[R(C/L)^{1/2} + G(L/C)^{1/2}] = \frac{p}{u} + \alpha = p\delta + \alpha
\end{aligned}$$

where $Z_c \approx R_c$ and $\delta = (LC)^{1/2}$. It then follows that

$$V(x, p) = e^{-\alpha x} V_1 e^{-p\delta x} + e^{\alpha x} V_2 e^{+p\delta x}$$

$$V(x, p) = e^{-\alpha x}\, M(p)\, e^{-p\delta x} + e^{-\alpha(2l-x)}\, N(p)\, e^{-p\delta(2l-x)}$$

$$v(x, t) = e^{-\alpha x} m(t - \delta x) \Upsilon(t - \delta x) + e^{-\alpha(2l-x)} n[t - \delta(2l - x)] \Upsilon[t - \delta(2l - x)]$$

where $\Upsilon(t)$ is the unit step function.

In the same way, since Z_c is approximately a pure resistance R_c,

$$i(x, t) = e^{-\alpha x} \frac{m(t - \delta x) \Upsilon(t - \delta x)}{R_c} - e^{-\alpha(2l-x)} \frac{n[t - \delta(2l - x)] \Upsilon[t - \delta(2l - x)]}{R_c}$$

Here $m(t)$ and $n(t)$ are functions depending on the source.

(b) Line without Dispersion ($R/L = G/C = a$). In this case

$$Z_c(t) = R_c = (L/C)^{1/2}$$

is independent of t, and where $\alpha = a\delta$,

$$\gamma(p) = (\mathrm{LC})^{1/2}(p + a) = (p + a)\delta = p\delta + \alpha$$

This is the same expression for $\gamma(p)$ as for the beginning of the transient regime. Thus the expressions for $v(x, t)$ and $i(x, t)$ obtained in Section 5.2(a) apply also here.

(c) Lossless Line R = G = 0, $a = 0$. Here

$$v(x, t) = m(t - \delta x)\Upsilon(t - \delta x) + n[t - \delta(2l - x)]\Upsilon[t - \delta(2l - x)]$$

$$i(x, t) = \frac{m(t - \delta x)\Upsilon(t - \delta x)}{\mathrm{R}_c} - \frac{n[t - \delta(2l - x)]\Upsilon[t - \delta(2l - x)]}{\mathrm{R}_c}$$

The step functions $\Upsilon(t - \delta x)$ and $\Upsilon[t - \delta(2l - x)]$ indicate that the relations above hold beginning at the instant that the waves reach the point x.

(d) Conclusions. In the preceding relations, a parameter appears, which has the following values:

$\alpha = \frac{1}{2}(\mathrm{R}/\mathrm{R}_c + \mathrm{GR}_c)$ in the beginning of the transient regime;
$\alpha = a\delta$ for a line without dispersion (Heaviside condition);
$\alpha = 0$ for a lossless line.

In these three special cases, the voltage and current have been decomposed into two systems of traveling waves, propagating with the same velocity $1/\delta = 1/(\mathrm{LC})^{1/2} = u$ (this uniformity of the propagation velocities is rigorous if the Heaviside condition is satisfied, and is well approximated at the beginning of the transient regime).

The first set of waves is

$$v_i = e^{-\alpha x} m(t - \delta x)\Upsilon(t - \delta x)$$

$$i_i = \frac{v_i}{\mathrm{R}_c}$$

These propagate in the direction of increasing x, and are thus the incident waves. The voltage and current at each instant are in the same ratio, the characteristic impedance $\mathrm{R}_c = (\mathrm{L}/\mathrm{C})^{1/2}$. Both waves decrease exponentially in the direction of propagation with an exponent, or attenuation constant, α.

The second set of waves is

$$v_r = e^{-\alpha(2l-x)} n[t - \delta(2l - x)]\Upsilon[t - \delta(2l - x)]$$

$$i_r = \frac{-v_r}{\mathrm{R}_c}$$

These propagate in the direction of decreasing x, and are thus reflected waves. At each instant the voltage and current have opposite phases; that is, their ratio is the negative of the characteristic impedance, $-R_c = -(L/C)^{1/2}$. They decrease exponentially in the direction of propagation with an attenuation constant α. The current wave and the voltage wave are of opposite sign.

Let us summarize the formulas upon which the method of traveling waves is based:

$$v = v_i + v_r$$

$$i = i_i + i_r; \qquad i_i = \frac{v_i}{R_c}; \qquad i_r = -\frac{v_r}{R_c}$$

$$v_i = R_c i_i; \qquad v_r = -R_c i_r; \qquad R_c = (L/C)^{1/2}$$

Exercises

The solutions to those exercises marked* will be found at the end of the book, in Chapter VI.

*EXERCISE 1. Consider two identical lossless lines, arranged as shown in Fig. 53. At the instant $t = 0$, the current source is removed by opening the switch. (Assume that prior to $t = 0$ the steady state had been reached.)

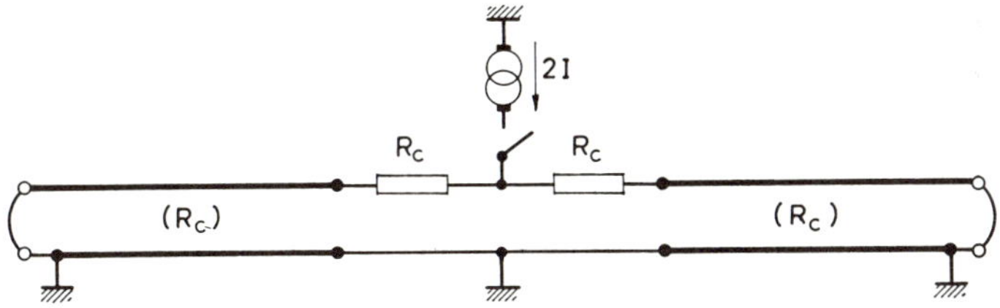

Fig. 53. Circuit for Exercise 1.

1. Study the transient regime, using the method of traveling waves.
2. Explain why the circuit does not come to rest, in spite of the presence of two dissipative elements R_c at the inputs to the lines.
3. Repeat the reasoning, this time taking account of losses in the line.

*EXERCISE 2. Consider a lossless line, open-circuited at the load end. Let R_c be the characteristic resistance. The line is excited at the input end by the circuit shown in Fig. 54. Assuming the diode is ideal, and has the characteristic shown in Fig. 55, find the voltages and currents at A and B.

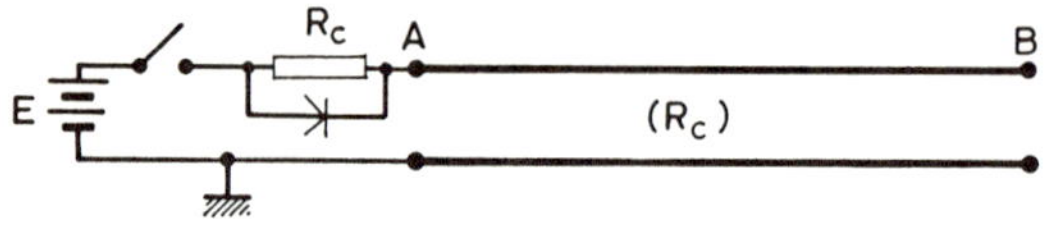

Fig. 54. Circuit for Exercise 2.

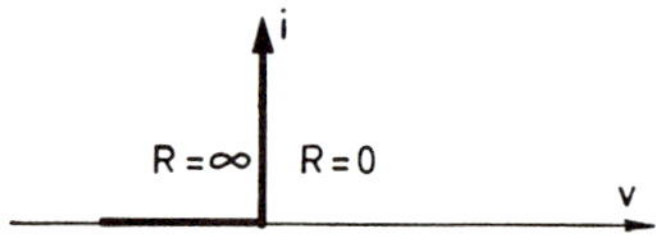

Fig. 55. Characteristic of diode in Fig. 54.

*EXERCISE 3. Demonstrate the result mentioned in Chapter II, according to which a self-inductance at the load end of a line can be matched using a second line terminated in a capacitance, in the arrangement shown in Fig. 56.

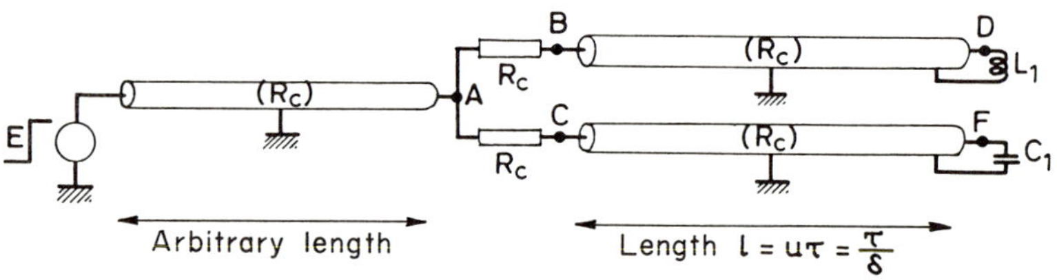

Fig. 56. Circuit for Exercise 3.

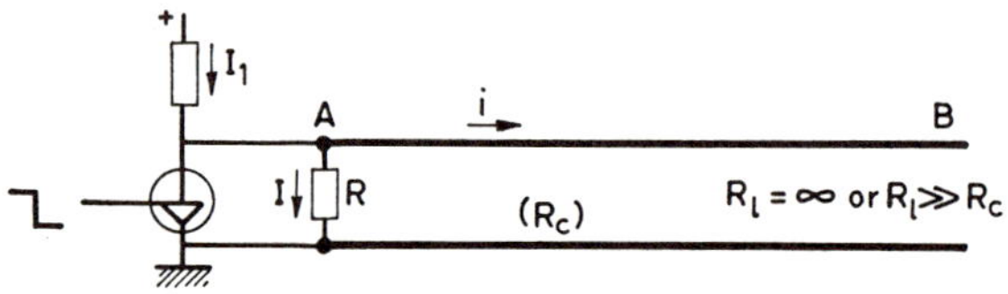

Fig. 57. Circuit for Exercise 4.

*EXERCISE 4 (Fig. 57). Prior to $t = 0$, the transistor is conducting and saturated, and the voltage on the line is essentially zero. At $t = 0$ the transistor is cut off, in a time $\ll \tau$. A current generator furnishes the current I_1.

Find the form of the voltages at A and at B, assuming first that R is much larger than R_c, and then that R is much smaller than R_c.

*EXERCISE 5. Consider the transmission of a 1-volt step through a conductor which has an ohmic resistance which is not negligible with respect to its characteristic resistance R_c. Assume that the source resistance is $R_0 = 10\,\Omega$, the characteristic resistance is $R_c = 50\,\Omega$, and the load resistance is $R_l = 100\,\Omega$. Suppose that the ohmic resistance of the conductor is $20\,\Omega$.

Determine the voltage at different points along the conductor by separating the resistive conductor into a number of elementary lumped resistances ΔR, joined by conductors with zero ohmic resistance.

As a first approximation, decompose the 20-Ω resistance into only two lumped elements ΔR. Assume that the lengths of the lumped resistors ΔR are negligible with respect to the lengths of the lines connecting them. Use the tabular method.

*EXERCISE 6. Consider an assemblage of three lines joined together as in Fig. 58. Using the tabular method, determine the voltages at points A, M, S_1, and S_2 after a step of voltage of 1 volt has been applied at point A.

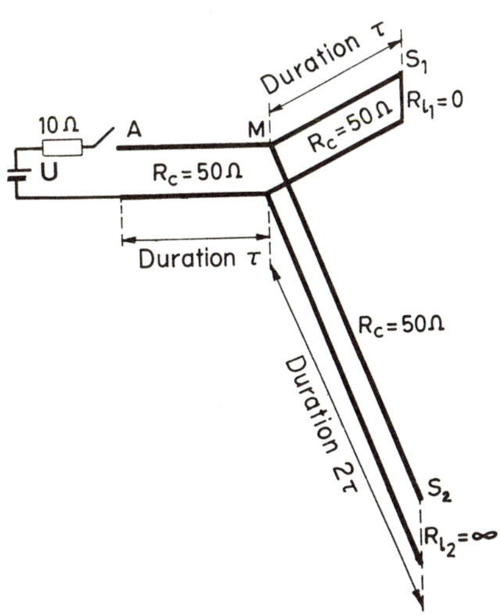

Fig. 58. Circuit for Exercise 6.

CHAPTER III

The Bergeron Diagram

1. Basic Relations

The method of Bergeron† is a graphical presentation of the method of traveling waves, just as the tabular method described in the preceding chapter is a numerical presentation of the same ideas. It is especially easy to use in the case of lossless lines, but it can be applied also, although less easily, to lines with losses. It has however the great advantage common to graphical methods in general, of being applicable to circuits in which the $v(i)$ characteristic is nonlinear. On the other hand, as we shall see, the graphs involved become quite complex if reactive elements are present.

We will need to use the relations for a lossless line, developed in Section 4.2 of Chapter I, which give the voltage v and current i at each time t and at each point x along the line:

$$\begin{aligned} v(x, t) &= g(x - ut) + h(x + ut) \\ i(x, t) &= \frac{1}{\mathrm{R}_c} [g(x - ut) - h(x + ut)] \end{aligned} \tag{III.1}$$

Here, as always, $\mathrm{R}_c = (\mathrm{L/C})^{1/2}$ is the characteristic resistance, and $u = 1/(\mathrm{LC})^{1/2}$ is the propagation velocity. The positive senses of the abscissa

† Bergeron, a French hydraulic engineer, developed this method to study the propagation of water hammer waves in hydraulics. See L. J. B. Bergeron, "Du Coup de Bélier en Hydraulique au Coup de Foudre en Électricité." Dunod, Paris, 1949. (English transl. "Water Hammer in Hydraulics and Wave Surges in Electricity." Wiley, New York, 1961.)

x, the propagation velocity, and the displacement along the line are all taken from the input towards the output.

Consider an observer O_1 traveling along the line with velocity $+u$, in the direction from input towards output, who is at the point $x = x_0$ at

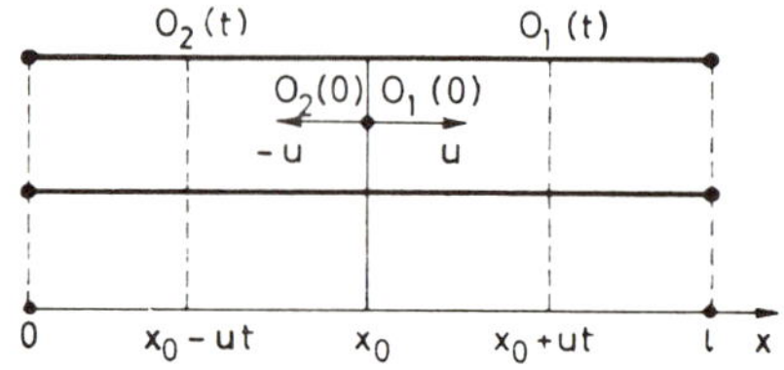

Fig. 59. Observers in Bergeron's method.

time $t = 0$ (Fig. 59). At that time and at that point he observes the voltage and current

$$v(x, 0) = v_0 = g(x_0) + h(x_0) \tag{III.2}$$

$$i(x, 0) = i_0 = \frac{1}{R_c} [g(x_0) - h(x_0)] \tag{III.3}$$

At some other time t, this observer will arrive at the point $x = x_0 + ut$, and will observe the voltage and current

$$\begin{aligned} v[(x_0 + ut), t] &= g[(x_0 + ut) - ut] + h[(x_0 + ut) + ut] \\ &= g(x_0) + h(x_0 + 2ut) \end{aligned}$$

$$\begin{aligned} i[(x_0 + ut), t] &= \frac{1}{R_c} \{[g(x_0 + ut) - ut] - h[(x_0 + ut) + ut]\} \\ &= \frac{1}{R_c} [g(x_0) - h(x_0 + 2ut)] \end{aligned}$$

During the time of travel t from x_0 to $x_0 + ut$, he has thus encountered the changes in voltage and current:

$$\Delta v_1 = v[(x_0 + ut), t] - v(x_0, 0) = h(x_0 + 2ut) - h(x_0) \tag{III.4}$$

$$\Delta i_1 = i[(x_0 + ut), t] - i(x_0, 0) = \frac{-1}{R_c} [h(x_0 + 2ut) - h(x_0)] \tag{III.5}$$

These changes are seen to be related by

$$\Delta v_1 = -R_c \, \Delta i_1 \tag{III.6}$$

Thus an observer moving from the input towards the output, observes voltage and current changes which are always in the ratio $-R_c$. This result can be expressed graphically as follows. Assume that the observer has seen at $t = 0$ a voltage and current v_0 and i_0, plotted as the point M_0 in Fig. 60. At some other arbitrary time t, he will observe a state (v, i) such that

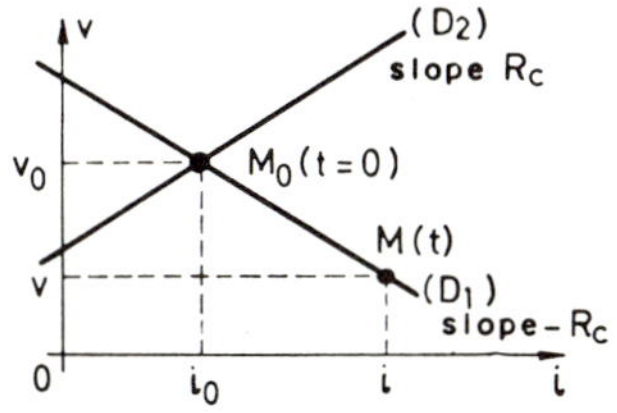

Fig. 60. Voltages seen by the observers of Fig. 59.

$\Delta v_1 = -R_c \Delta i_1$. The locus of such states is the line D_1, which has slope $-R_c$ and passes through the point M_0.

Consider now a second observer, who moves along the line with velocity $-u$, and who at $t = 0$ is also at x_0 (Fig. 59). At that point and time, he observes

$$v(x_0, 0) = v_0 = g(x_0) + h(x_0)$$

$$i(x_0, 0) = i_0 = \frac{1}{R_c}[g(x_0) - h(x_0)]$$

At some other time t, he will have arrived at the point $x_0 - ut$, and will observe there

$$v[(x_0 - ut), t] = g(x_0 - 2ut) + h(x_0)$$

$$i[(x_0 - ut), t] = \frac{1}{R_c}[g(x_0 - 2ut) - h(x_0)]$$

Thus during this displacement, he will have encountered a change in voltage and current

$$\Delta v_2 = g(x_0 - 2ut) - g(x_0) \tag{III.7}$$

$$\Delta i_2 = \frac{1}{R_c}[g(x_0 - 2ut) - g(x_0)] \tag{III.8}$$

These two changes are thus related by

$$\Delta v_2 = R_c \Delta i_2 \tag{III.9}$$

Hence this observer, who moves from the output towards the input, sees variations in voltage and current which are in the ratio $+R_c$. The voltages and currents encountered by this observer lie along the line of slope $+R_c$ passing through M_0 in Fig. 60.

It is especially important to note that a point on one of these lines has as coordinates the true voltage and current encountered by the observer at some point on the line, and not the changes in these quantities. In Fig. 61

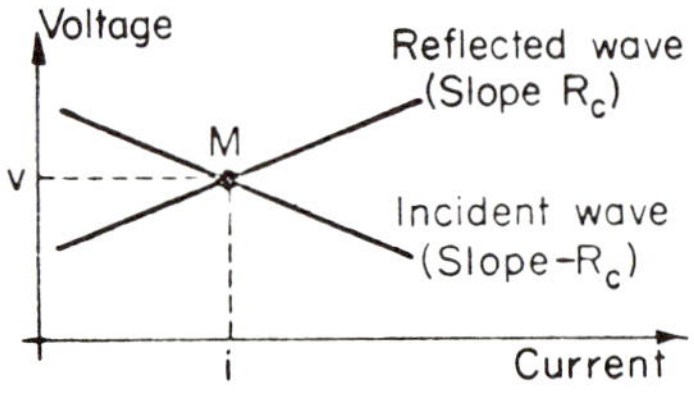

Fig. 61. Voltages seen by general observers.

are shown these two lines, called characteristic lines of the two waves, incident and reflected, as they would be drawn considering some point on the line and some instant as origin. Here again the total voltage v and current i are plotted.

2. Principle of the Method of Bergeron

2.1 *The method using a single observer* (Fig. 62)

Let us study affairs at the two points A and B of a line, knowing that the voltages and currents at these points are related by

$$v_A = F(i_A); \qquad v_B = G(i_B)$$

Here F and G, the input and output characteristics of the line, may be functions of time.

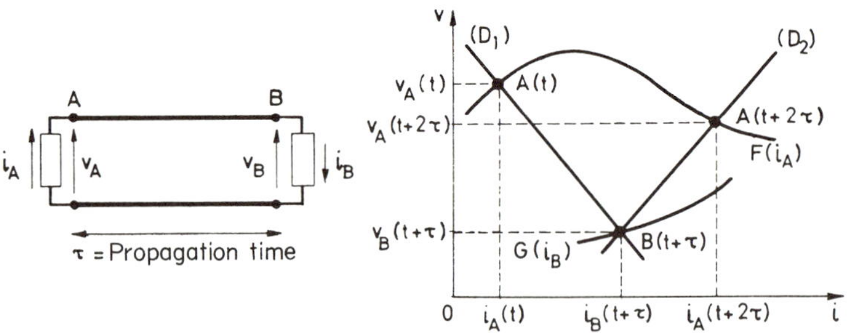

Fig. 62. Sign conventions, and the state diagram in Bergeron's method.

An observer leaving A at time t with velocity $+u$ sees states (v, i) along the line which have as their locus the line D_1 in Fig. 62, with slope $-R_c$ and passing through the point $A(t)$. At time $t + \tau$, he arrives at point B, where the state of the transmission line is represented by the point $B(t + \tau)$, the intersection of the characteristic D_1 and the curve $v_B = G(i_B)$. At this instant $t + \tau$, the observer leaves B and returns to A with velocity $-u$. He then encounters states on the line D_2, with slope $+R_c$ and passing through $B(t + \tau)$. At time $t + 2\tau$ he arrives at A, where the state of the transmission line is represented by the point $A(t + 2\tau)$, the intersection of the line D_2 and the curve $v_A = F(i_A)$.

Continuing this process back and forth across the transmission line, the successive states at A and B can be found. If the curves F and G happen to intersect at some point, this is the steady state of the line, and the graphical process stops there. Otherwise, the process continues indefinitely as an oscillation between two stable states.

All this will be made more precise by means of examples.

1. *First example. Line terminated in a resistance* R_l *and excited by a voltage step at the input* (Fig. 63)

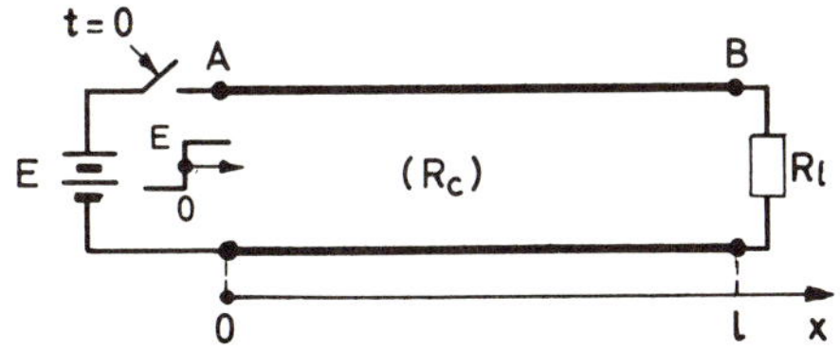

Fig. 63. Line charged by a voltage step.

At the instant $t = 0$, the switch is closed and the voltage E is applied to the line, producing a voltage step at the input.

On the (v, i) plot, the loci defining the conditions imposed at the ends of the line A and B by the external elements are plotted. These are a line C_0 parallel to the i axis and with ordinate E (valid for $t \geq 0$), and a line C_l of slope R_l passing through the origin (valid for all t).

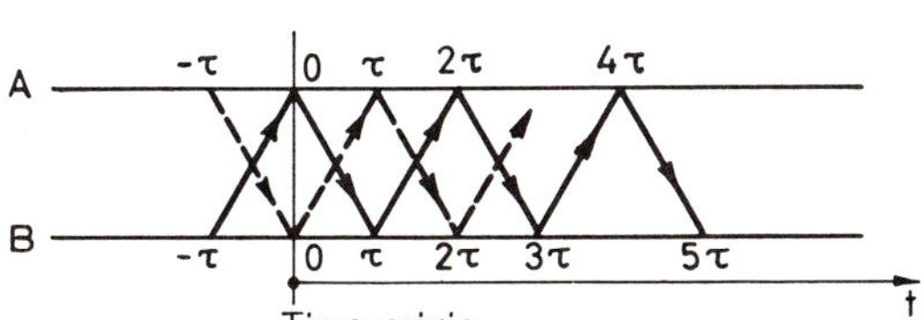

Fig. 64. Timing chart for circuit of Fig. 63.

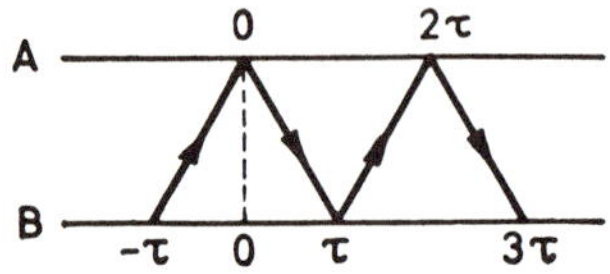

Fig. 65. Timing diagram for first observer in circuit of Fig. 63.

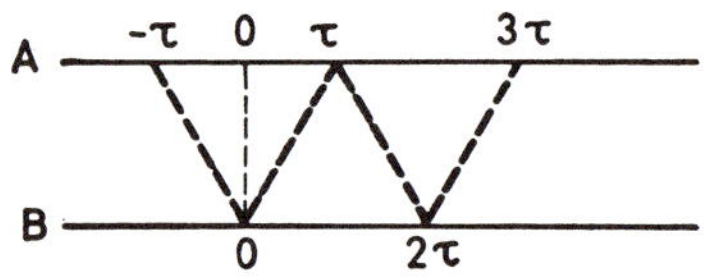

Fig. 66. Timing diagram for second observer in circuit of Fig. 63.

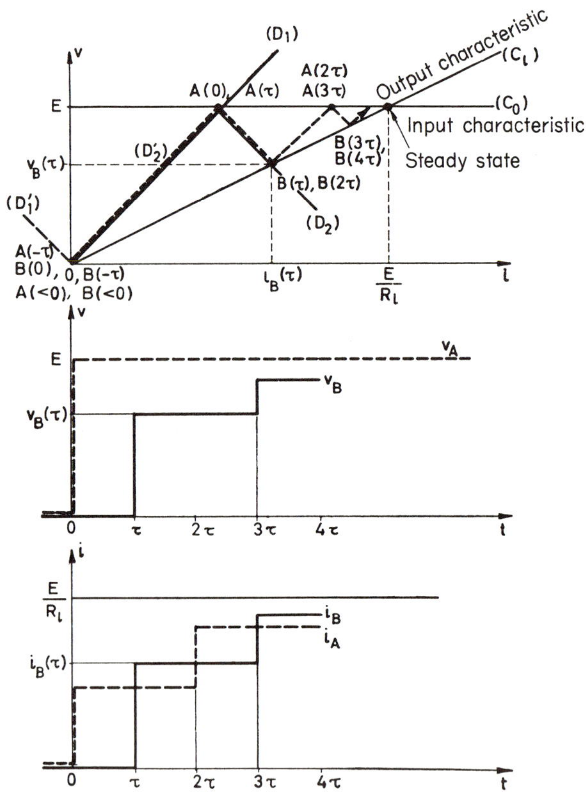

Fig. 67. Bergeron diagram for circuit of Fig. 63, for $R_l < R_c$.

In addition, a second graph of the type used in railway operation is set up (Fig. 64). This provides a means of knowing the instants at which the observer reaches A and B as a function of the time he first departs from A. In considering many to-and-fro journeys of the observer, it may be necessary to determine the states at A and B only at certain times of particular interest.

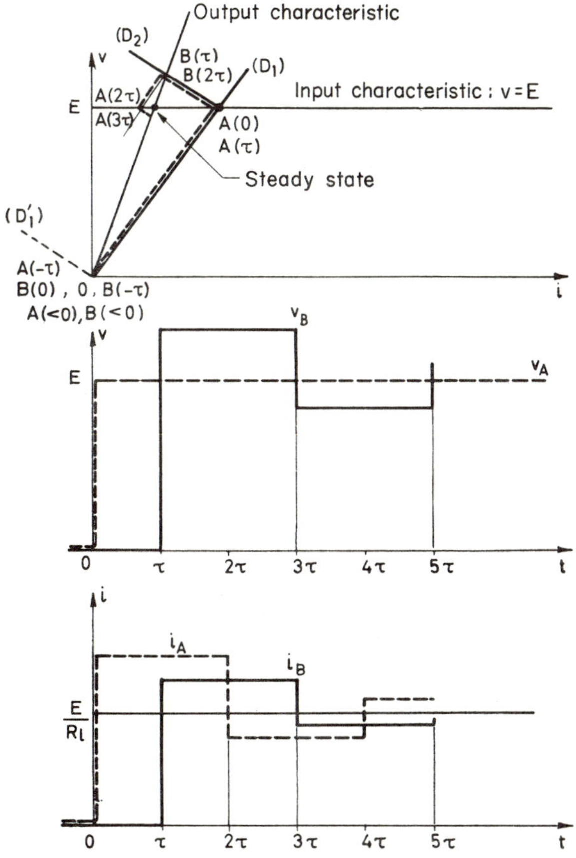

Fig. 68. Bergeron diagram for circuit of Fig. 63, for $R_l > R_c$.

In Figs. 67, 68, and 69, the characteristic lines have been drawn with a slope of 45 deg, which simply amounts to assuming certain scales for v and i.

For times $t < 0$, the state of the line is at the origin. A(< 0) and B(< 0) are written at this point to indicate that this is the state of A and B for $t < 0$.

In order to determine the evolution of a phenomenon beginning at $t = 0$, at point A, an observer is caused to depart from point B at $t = -\tau$. He arrives at A at $t = 0$, after encountering states lying on the line D_1 (Figs. 67–69). The points $B(-\tau)$ and $A(0)$ are placed on the graphs appropriately to indicate the states at those points at times $-\tau$ and 0.

This observer then departs for B, where he arrives at $t = \tau$ (Fig. 65). During this journey he encounters states lying on the line D_2. The point $B(\tau)$ represents the state of B at time τ.

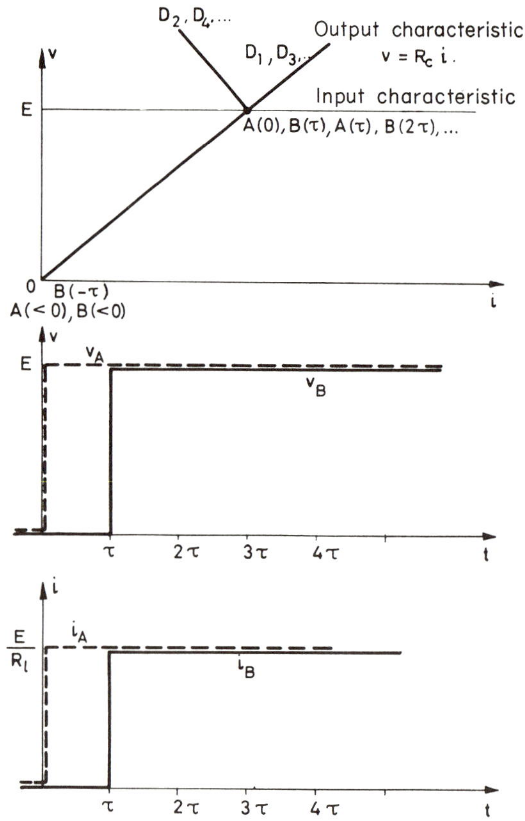

Fig. 69. Bergeron diagram for circuit of Fig. 63, for $R_l = R_c$.

Continuing this process, $A(2\tau)$, $B(3\tau)$, etc. are obtained. After infinitely many journeys, our traveler would reach the intersection of the lines C_0 and C_l, which is the steady state at each point of the line: $v = E$, $i = E/R_l$.

Once the graph is completed, the voltages and currents at various times are read off, and v_A, i_A, v_B, i_B thus obtained as time functions.

It remains to determine the state at B at the times 0, 2τ, 4τ, ..., and at A at τ, 3τ, 5τ, ... These can be found by applying the method of traveling waves of Chapter II. But they can also be recovered from the graph by considering a second voyage of the observer (Fig. 66). He leaves A at a time $-\varepsilon$, which for convenience will be taken as $-\tau$. The lines D_1', D_2', ... and the states $A(-\tau)$, $B(0)$, $A(\tau)$, $B(2\tau)$, $A(3\tau)$, ... are then obtained (Figs. 67–69).

Of the three cases illustrated in Figs. 67–69, $R_l < R_c$, $R_l = R_c$, and $R_l > R_c$, the importance of the matched line ($R_l = R_c$) is clear.

2. *Second example. Nonlinear problems*

It is not necessary that the voltage-current characteristics of the generator feeding the line at A, and of the load at B, be straight lines as they were in the first example. It is one of the advantages of Bergeron's method that nonlinear characteristics can be dealt with easily.

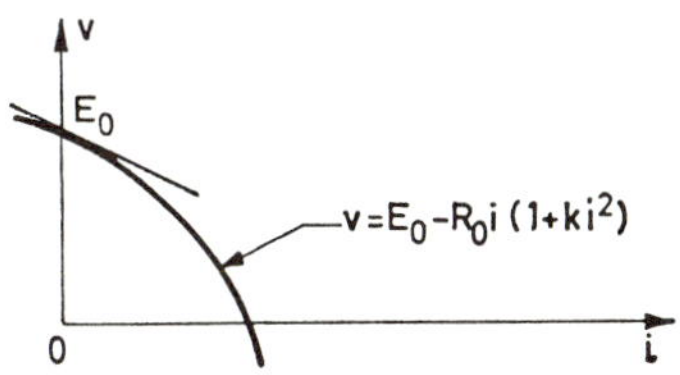

Fig. 70. Volt-ampere characteristic of a generator.

Consider for example a line fed at A by a generator having a characteristic as shown in Fig. 70. This could be a generator in which the internal resistance is a function of temperature, and thus of current:

$$R = R_0(1 + \alpha\theta) = R_0(1 + ki^2)$$

Suppose further that the line is loaded at B by a diode having the characteristic shown in Fig. 71. This problem has the graphical solution indicated in Fig. 72.

Fig. 71. Volt-ampere characteristic of a nonlinear load.

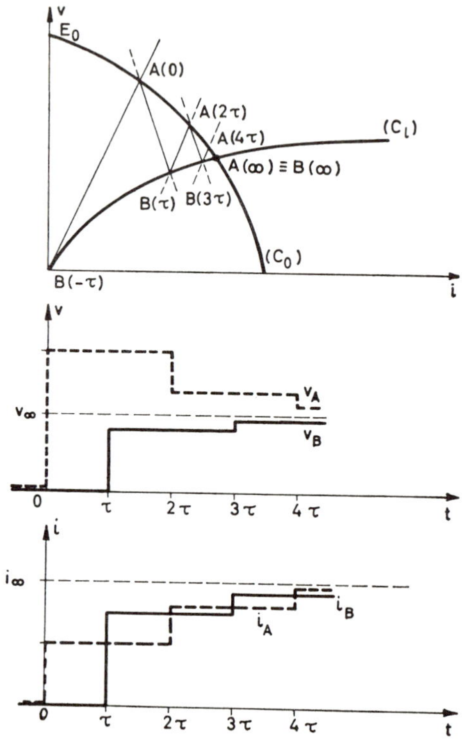

Fig. 72. Bergeron diagram and waveforms resulting from the source of Fig. 70 and the load of Fig. 71.

2.2 *The method of simultaneous observers*

1. *Principle*

This method can be used to solve the following problem: Knowing the voltage and current at each instant at each end of the line A and B, find the voltage and current at each instant at any arbitrary point M along the line (Fig. 73). The voltages and currents at A and B are assumed to have been found already using the method of a single observer, discussed in the preceding section.

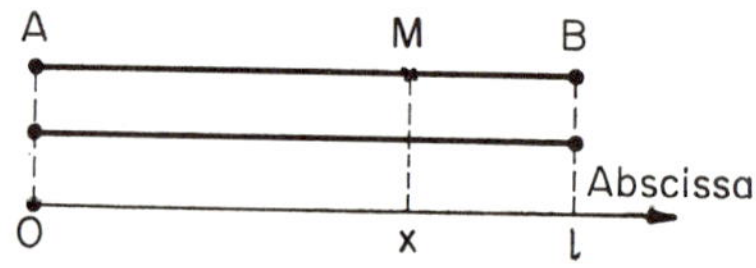

Fig. 73. A general point on a line.

The method of simultaneous observers consists in starting two observers O_1 and O_2 from A and B, respectively, at suitable instants, traveling towards each other, and arriving together at the point M of interest at the time t at which the state at M is to be found.

Let

$$\theta_1 = \frac{x}{u}; \qquad \theta_2 = \frac{l - x}{u}$$

be the times of propagation across AM and MB, respectively. A "railroad" graph, Fig. 74, gives the instants at which the two observers must depart.

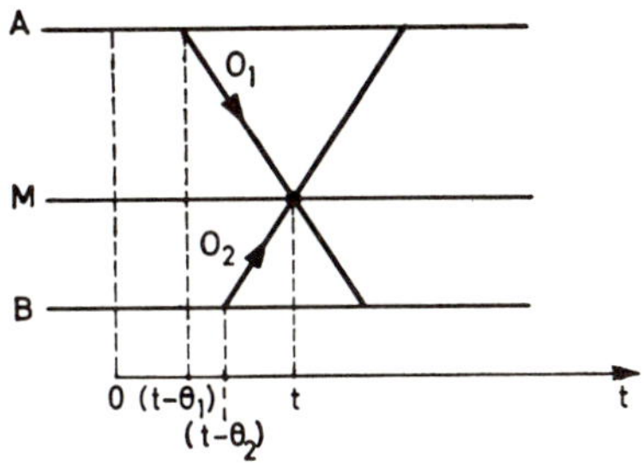

Fig. 74. Timing chart for simultaneous observers on the line of Fig. 73.

Now consider the (v, i) diagram showing the state of every point on the line as a function of time (Fig. 75). Let $A(t - \theta_1)$ be the point representing the state at the input to the line at time $t - \theta_1$, and $B(t - \theta_2)$ the point representing the output state at $t - \theta_2$. Observer O_1 leaves A at time $t - \theta_1$ and travels with velocity $+u$. The locus of states which he encounters is a

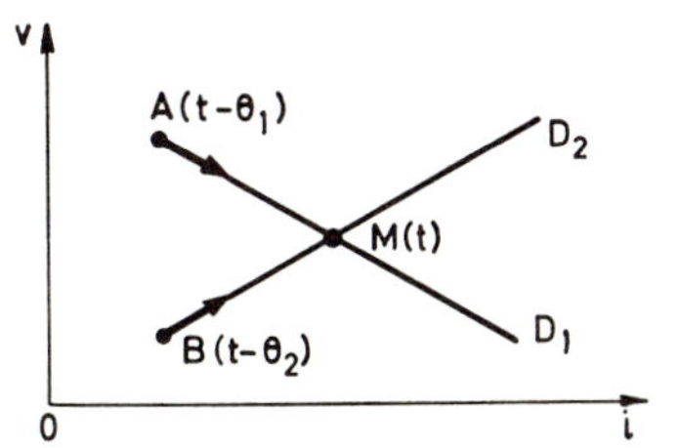

Fig. 75. States encountered by the observers of Fig. 74.

straight line D_1, with slope $-R_c$ and passing through $A(t - \theta_1)$. Observer O_2 leaves B at $t - \theta_2$ and travels with velocity $-u$. He encounters states lying on the straight line D_2, having slope $+R_c$ and passing through $B(t - \theta_2)$.

At the time t of interest, the two observers meet at the point M along the line of interest. The two observers then being at the same place on the line at the same time must see the same state, which is hence the intersection of the lines D_1 and D_2 in Fig. 75. This point of intersection is thus the desired state $M(t)$, representing the voltage and current at M at time t.

In the case of a line excited by a step function of voltage, the voltage or the current or both change their value at point A at times $0, 2\tau, 4\tau, \ldots, 2k\tau, \ldots$, and at point B at times $\tau, 3\tau, 5\tau, \ldots, (2k + 1)\tau, \ldots$. Hence the state of point M changes at the times $2k\tau + \theta_1$ because of changes at A, and at the times $(2k + 1)\tau + \theta_2$ because of changes at B. Further, the states at A, B, and M are related by

$$M(\theta_1 + 2k\tau) \equiv A(2k\tau)$$

$$M(\theta_2 + (2k + 1)\tau) \equiv B((2k + 1)\tau)$$

Two examples will serve to illustrate these statements.

2. *First example*

Consider a voltage generator of voltage U and negligible internal resistance, exciting a lossless line having characteristic resistance R_c, loaded by a resistance $R_l = 2R_c$ (Fig. 76). We wish to find the voltage and current

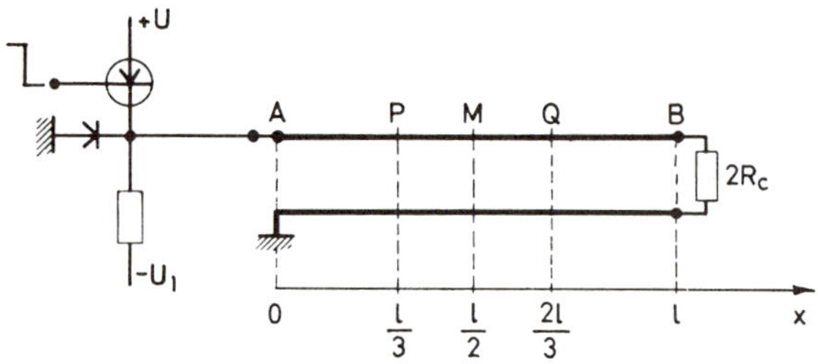

Fig. 76. Line charged by a voltage step.

waves at the input $A(x = 0)$, the output $B(x = l)$, the midpoint $M(x = l/2)$, a point $P(x = l/3)$, and a point $Q(x = 2l/3)$. As always, let τ be the time of propagation across the entire line.

(a) THE STATE GRAPH (Fig. 77). At times $t < 0$ nothing has yet occurred, and the state of the line is at the origin of the state diagram, $v = i = 0$. At $t = 0$, the transistor is turned on, and immediately saturates. An observer leaving B at time $-\tau$ (at which $v = i = 0$), arrives at A at $t = 0$ and encounters $v = U$, $i = U/R_c$. The state thus moves instantaneously from the origin to the point A(0) in Fig. 77.

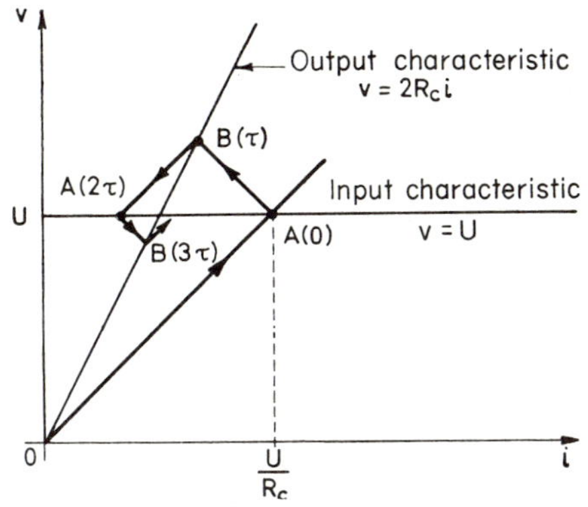

Fig. 77. Bergeron diagram for the line of Fig. 76.

A second observer leaves A at time 0, and travels towards B, where he arrives at time τ. The representative point moves from A(0) to B(τ) along a line with slope $-R_c$, B(τ) being the intersection of this line with the output characteristic.

The remainder of the graph can easily be constructed.

(b) THE VOLTAGE AND CURRENT AT A. At time $t = 0$ [the point A(0)], $v = U$, $i = U/R_c$. This state remains fixed until time 2τ [point A(2τ)], at which time the wave injected at A at time 0, and reflected at B, returns to A. This second state remains fixed until time 4τ [point A(4τ)], etc.

(c) THE VOLTAGE AND CURRENT AT B. Nothing occurs until time τ, at which time the wave leaving A at time 0 arrives at B, and the state moves to the point B(τ). This state remains fixed until time 3τ, at which time the wave reflected from A at time 2τ arrives at B and the state moves to the point B(3τ), etc.

(d) THE VOLTAGE AND CURRENT AT M (Fig. 78). Following the method explained previously, observers are allowed to depart from A and B at

Times $\left(t - \frac{\tau}{2}\right)$ at A	Times $\left(t - \frac{\tau}{2}\right)$ at B	Times t at M	State at M
-0.5τ	-0.5τ	0	$v = 0 \quad i = 0$
0		0.5τ	A(0)
	τ	1.5τ	B(τ)
2τ		2.5τ	A(2τ)
	3τ	3.5τ	B(3τ)
4τ		4.5τ	A(4τ)
	5τ	5.5τ	B(5τ)

Fig. 78. States at A and B influencing state at M in circuit of Fig. 76.

Times $(t-\frac{\tau}{3})$ at A	Times $(t-\frac{2\tau}{3})$ at B	Times t at P	State at P
$-\frac{\tau}{3}$	$-\frac{2\tau}{3}$	0	$v=0,\ i=0$
0		$\frac{\tau}{3}$	A(0)
	τ	$\frac{5\tau}{3}$	B(τ)
2τ		$\frac{7\tau}{3}$	A(2τ)
	3τ	$\frac{11\tau}{3}$	B(3τ)
4τ		$\frac{13\tau}{3}$	A(4τ)

Fig. 79. States at A and B influencing state at P in circuit of Fig. 76.

times such that they arrive together at M. The table of Fig. 78 is then constructed. An entry is made for each time at which the current, or the voltage, or both at the same time, changes at either A or B. The table is filled in, remembering that changes at A or B at times $t - \tau/2$ arrive at M at time t. The third column of the table shows the times at which changes in the state of M occur.

(e) VOLTAGE AND CURRENT AT P (Fig. 79). In order to determine the

Times $(t-\frac{2\tau}{3})$ at A	Times $(t-\frac{\tau}{3})$ at B	Times t at Q	State at Q
$-\frac{2\tau}{3}$	$-\frac{\tau}{3}$	0	$v=0,\ i=0$
0		$\frac{2\tau}{3}$	A(0)
	τ	$\frac{4\tau}{3}$	B(τ)
2τ		$\frac{8\tau}{3}$	A(2τ)
	3τ	$\frac{10\tau}{3}$	B(3τ)
4τ		$\frac{14\tau}{3}$	A(4τ)

Fig. 80. States at A and B influencing state at Q in circuit of Fig. 76.

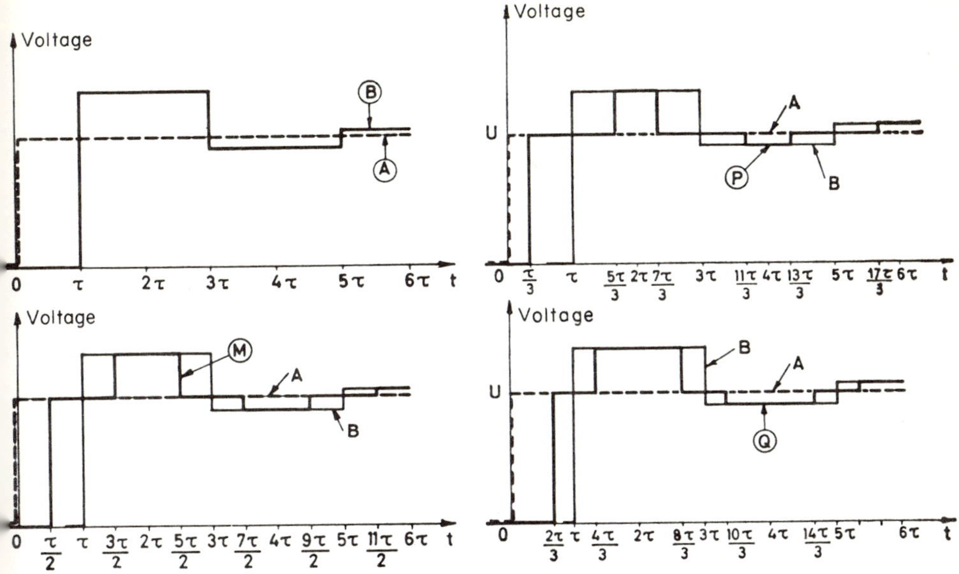

Fig. 81. Voltage waveforms in circuit of Fig. 76.

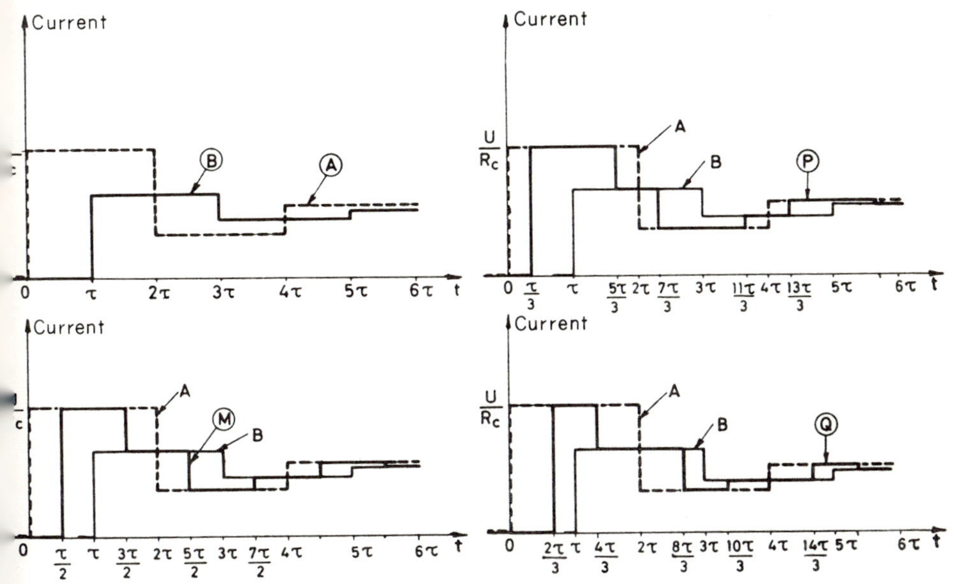

Fig. 82. Current waveforms in circuit of Fig. 76.

Experimental results (Tektronix 581 oscilloscope):

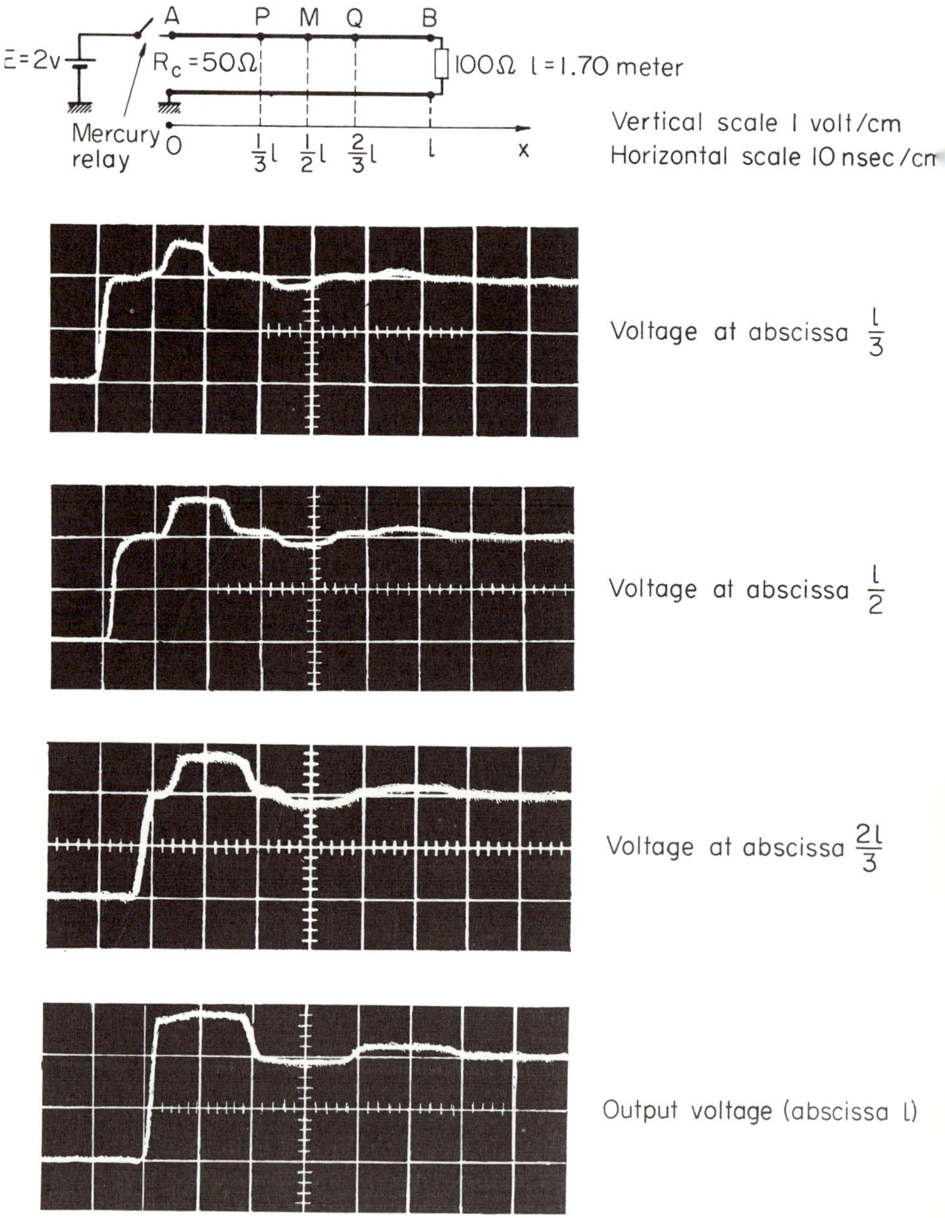

Fig. 83. Oscillograms of voltages in circuit of Fig. 76

state of point P at some time t, observers are sent out from A at time $t - \tau/3$, and from B at time $t - 2\tau/3$. The table of Fig. 79 may easily be constructed.

(f) Voltage and Current at Q (Fig. 80). In this case, observers leave A at $t - 2\tau/3$, and B at $t - \tau/3$. The table of Fig. 80 results.

All these results are shown in Figs. 81 and 82. Actual oscillographs, taken on a line of length 1.7 meters, are shown in Fig. 83.

3. *Second example. Junction of lines having different characteristic impedances* (Fig. 84)

This problem has already been discussed in Chapter II, and we will return to it again in Chapter IV. It is sometimes instructive to study the same problem using a number of different methods.

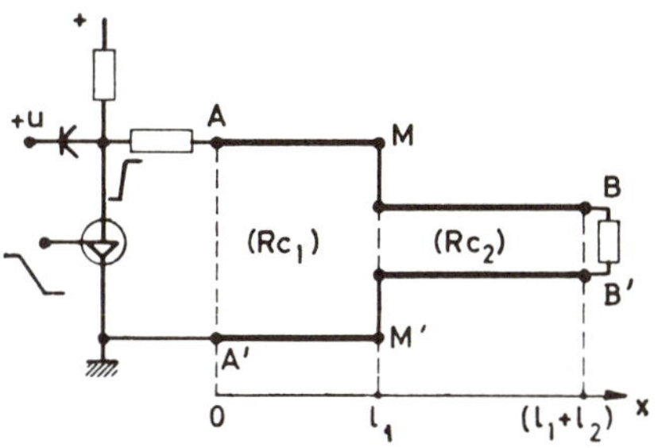

Fig. 84. Line with discontinuous characteristic resistance.

Let us consider two line segments, having different characteristic resistances R_{c1} and R_{c2} and lengths l_1 and l_2, respectively, and being joined end to end at (M, M′) (Fig. 84). A source is placed at the input (A, A′) of this composite line, and a resistance load at the output (B, B′). We wish to determine the state of the line at the points (A, A′), (B, B′), (M, M′). The time required to cut off the transistor is assumed to be much less than the propagation time τ.

Let u_1 and u_2 be the propagation velocities in the lines with characteristic resistances R_{c1} and R_{c2}, respectively. The propagation times through the line segments will then be, respectively, $\tau_1 = l_1/u_1$ and $\tau_2 = l_2/u_2$.

This problem is solved by considering the travels of two groups of observers. The observers of the first group travel between A and M in time τ_1, encountering states represented by points in the (v, i) plane on straight lines of slope $\pm R_{c1}$. The observers of the second group travel between points B and M in time τ_2, encountering states on lines with slopes $\pm R_{c2}$.

Let us find the state at point M at time t. Observers from these two groups leave A and B at times such that each observer from the first group meets an observer from the second group at M at time t, and vice versa. The "railway" graph of Fig. 85 shows the times at which observers O_1 and O_2 depart.

The state $A(t - \tau_1)$ at the input and the state $B(t - \tau_2)$ at the output

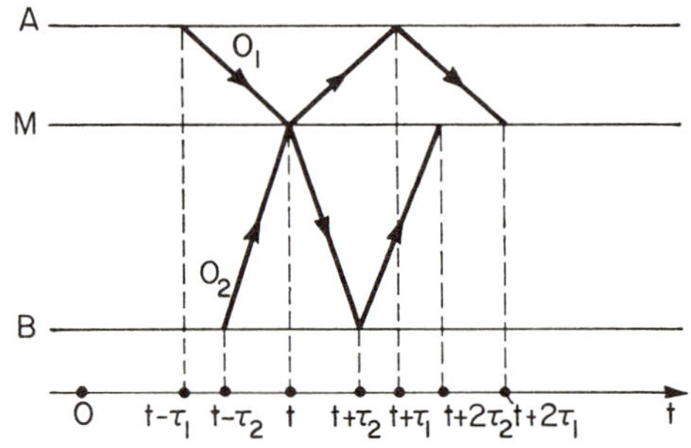

Fig. 85. Time chart for observers in circuit of Fig. 84.

are assumed known. An observer is caused to leave A at time $t - \tau_1$, and another to leave B at $t - \tau_2$. They meet at M at time t, and the state $M(t)$ is given by the intersection of the line with slope $-R_{c1}$ passing through $A(t - \tau_1)$ and the line with slope $+R_{c2}$ passing through $B(t - \tau_2)$ (Fig. 86).

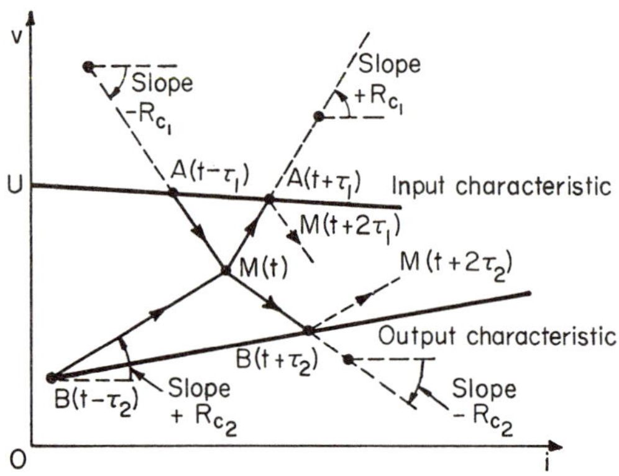

Fig. 86. First part of Bergeron's diagram for circuit of Fig. 84.

The first observer then retraces his path, and arrives at A at time $t + \tau_1$. The point $A(t + \tau_1)$ is at the intersection of the line through $M(t)$ with slope $+R_{c1}$, and the generator characteristic. This observer then departs again for M, where he arrives at time $t + 2\tau_1$. In the same way, the second observer leaves for B, which he reaches at $t + \tau_2$. The point $B(t + \tau_2)$ is the intersection of the line through $M(t)$ with slope $-R_{c2}$ and the characteristic curve of the output load resistance R_l. This observer then leaves again for M, which he reaches at $t + 2\tau_2$.

It is clear that, although this method is simple in principle, in practice its use is limited to cases in which the times τ_1 and τ_2 are in the ratio of small integers. In fact, it can be shown that if $\tau_1/\tau_2 = p/q$ it is necessary to consider $p + q$ observers, p leaving from A and q from B, in order that each observer leaving from A meet at M an observer from B.

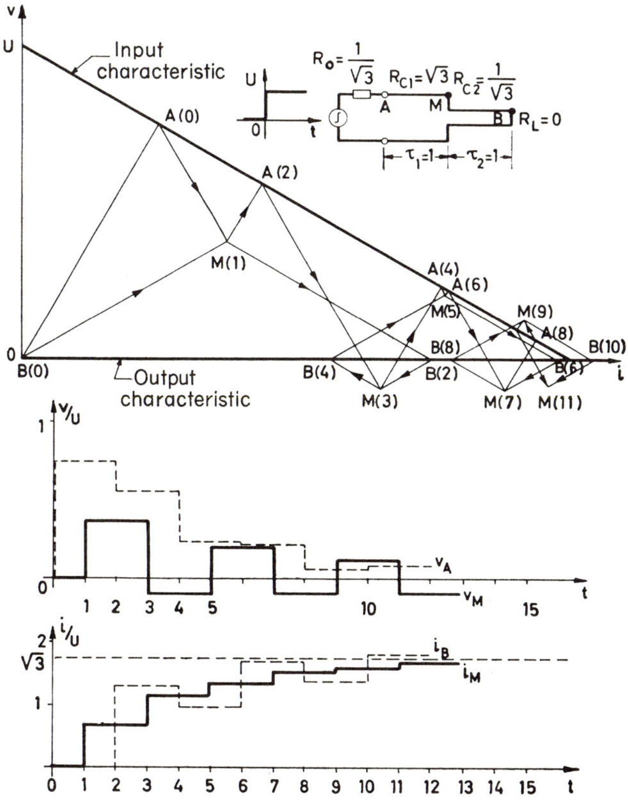

Fig. 87. Bergeron diagram for one case of the circuit of Fig. 84.

As an example, Fig. 87 shows the case of a line with characteristic resistance $3^{1/2}$ linked to a line with characteristic resistance $1/3^{1/2}$, with the latter terminated by a short circuit. The times τ_1 and τ_2 are assumed to be equal. The composite line is excited by a voltage step, furnished by a generator having an internal resistance of $1/3^{1/2}$. The times of transit being equal, it is sufficient to consider only two observers, one leaving A and one leaving B.

3. The Introduction of Lumped Elements

The method of Bergeron, as it has been described above, only allows distributed parameter circuits to be considered. This is insufficient, however, since in many problems of interest lumped elements are also present. The diagrams which result in the latter case are quite complicated, but we will nevertheless give some discussion of the subject.†

3.1 *Series resistance in the line* (Fig. 88)

This arrangement can be used to represent a line with loss. The states at M and N are to be found. Let θ_1 and θ_2 be the travel times from A to M and from N to B, respectively. The resistance R_1 being lumped (and of

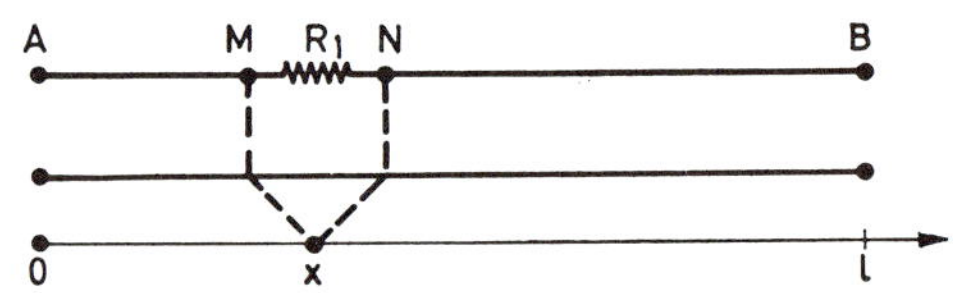

Fig. 88. A line with a lumped resistance inserted.

dimensions assumed negligible with respect to the length of the line), the propagation time from M to N is zero. The voltages and currents at A and B are assumed known at every instant t.

An observer O_1 leaves A at time $t - \theta_1$ and travels towards M, and observer O_2 leaves B at $t - \theta_2$ and travels towards N. They arrive at their destinations M and N at the same time t (Fig. 89). The points representing

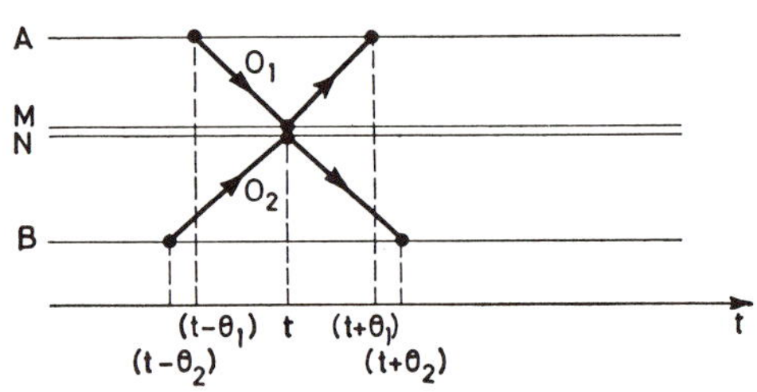

Fig. 89. Time chart for observers in circuit of Fig. 88.

the states seen by these observers O_1 and O_2 lie, respectively, on line D_1, having slope $-R_c$ and passing through the point $A(t - \theta_1)$, and on line D_2 with slope $+R_c$ and passing through $B(t - \theta_2)$ (Fig. 90).

† Bergeron, in his book (see footnote p. 65), considers at length circuits involving lumped resistance, self-inductance, and capacitance.

The resistance R_1 introduced between M and N forces the constraint $v_M - v_N = R_1 i$, represented in Fig. 90 by the line through the origin with slope R_1. The voltages corresponding to the states at M and N thus differ by $R_1 i$, while the currents at M and N are the same. In the state diagram,

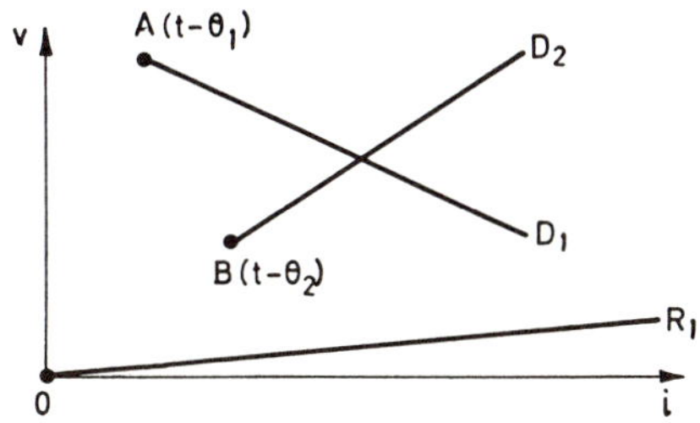

Fig. 90. States encountered by observers of Fig. 89.

the resistance R_1 thus leads to a vertical line connecting M(t) and N(t), the length of which is the ordinate to the line OR_1 at the abscissa i (Fig. 91).

The procedure is this (Fig. 91). At time $t - \theta_1$, a wave starts from A towards B. At $t - \theta_2$, a wave starts from B towards A. These two waves meet at time t, but on the graph the corresponding points at that time are

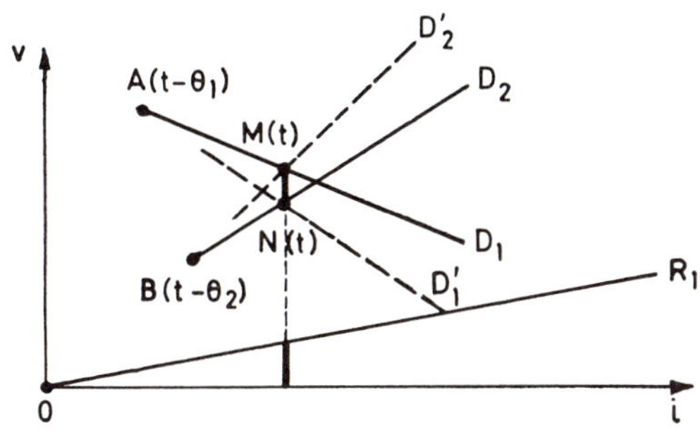

Fig. 91. States in circuit of Fig. 88.

M(t) and N(t). Next, the line D_1' is drawn, by subtracting the ordinates of line OR_1 from those of line D_1. Similarly, line D_2' results by adding the ordinates of line OR_1 to those of line D_2. M(t) and N(t) may then be found as the intersections of D_1 and D_2', and of D_2 and D_1', respectively.

3.2 *Parallel resistance across the line* (Fig. 92)

Here R_2 might represent a load fed by the line, a second load being at the end of the line (at B). The reasoning is the same as in the preceding paragraph, and the same notation will be used.

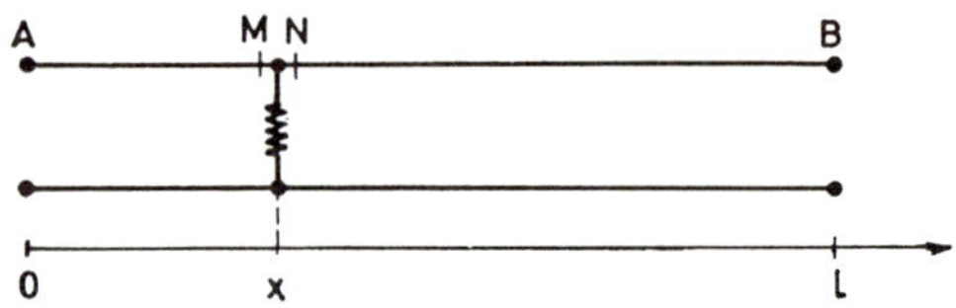

Fig. 92. A line with lumped resistance in shunt.

The states M(t) and N(t) in this case differ in their currents, while the voltages are the same. The current difference is the abscissa of the line OR_2. The construction of Fig. 93 follows.

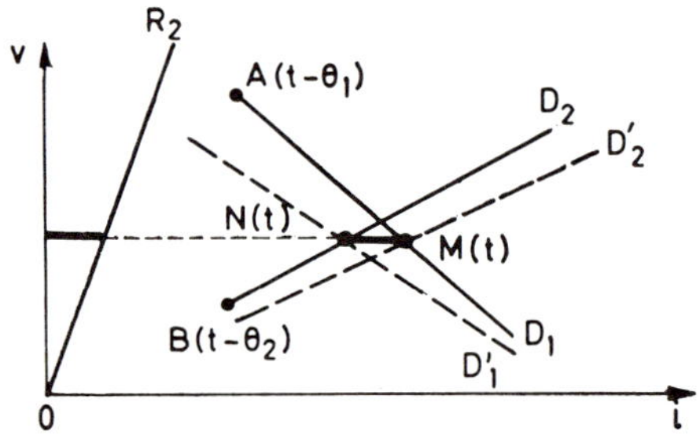

Fig. 93. Bergeron diagram for the circuit of Fig. 92.

3.3 *The combination of series and shunt resistance* (Fig. 94)

Because of the waves reflected at M and M″, the method, although simple in principle, in practice will be limited to the case that θ_1 and θ_2 are in the ratio of small integers. This comment, made already in Section 2.2,3, holds also for the cases of Sections 3.1 and 3.2.

Figure 94 is self-explanatory.

3.4 *Line terminated in an inductance* (Fig. 95)

The voltage at the terminals of a self-inductance is

$$v = L_1 \frac{di}{dt}$$

This relation will be approximated by the finite-difference equation

$$v = L_1 \frac{\Delta i}{\Delta t}$$

This approximation becomes better the smaller is Δt. But the smaller Δt

is taken, the more complex will be the graph to be constructed. For convenience, we will assume $\Delta t = \tau/n$, where n is an appropriate integer, and τ is the time of propagation through the line.

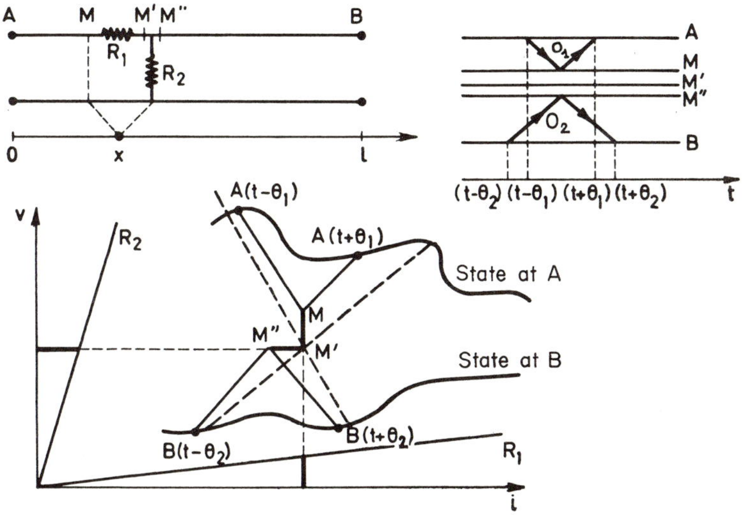

Fig. 94. Circuit with series and shunt lumped resistance, with time chart for observers and Bergeron diagram.

If the voltage at B is known at time t, and is to be found at time $t + \Delta t$, the average voltage at B can be approximated by

$$\frac{v_t + v_{t+\Delta t}}{2} = v = L_1 \frac{\Delta i}{\Delta t}$$

or

$$v = \frac{L_1}{\Delta t} \Delta i$$

Thus v and Δi are linearly related. Let us define

$$\tan \alpha = \frac{L_1}{\Delta t/2} = \frac{2L_1}{\Delta t}$$

Fig. 95. Line with inductive load.

The graph of Fig. 96 is constructed as follows. From the point B(t) the line making the angle $-\alpha$ with the i axis is drawn. This line cuts the i axis at some point K. Then from K the line making an angle $+\alpha$ with the i axis is drawn. The point B($t + \Delta t$) is then determined as the intersection of this latter line with the line representing an observer who departs from A at time $t + \Delta t - \tau$.

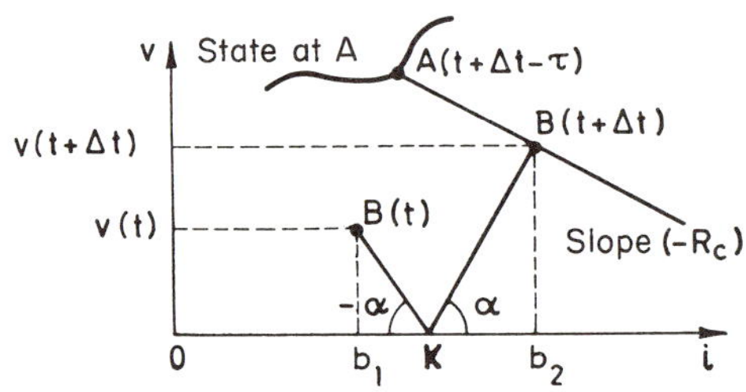

Fig. 96. Portion of state diagram for circuit of Fig. 95.

To justify this procedure, note that

$$\tan \alpha = \frac{v_t}{b_1 \mathrm{K}} = \frac{v_{t+\Delta t}}{\mathrm{K} b_2} = \frac{v_t + v_{t+\Delta t}}{b_1 b_2} = \frac{2v}{\Delta i}$$

so that, since $\tan \alpha = 2\mathrm{L}_1/\Delta t$,

$$\frac{2\mathrm{L}_1}{\Delta t} = \frac{2v}{\Delta i}$$

or

$$v = \mathrm{L}_1 \frac{\Delta i}{\Delta t}$$

which is the relation mentioned above.

It is of interest to remark that the diagram of Fig. 96 is the same as that which would result from a fictitious line of characteristic resistance $\mathrm{R}_c = 2\mathrm{L}_1/\Delta t$ and propagation time $\Delta t/2$. This representation of L_1 by a transmission line corresponds to a result to be developed in Section 6 of Chapter IV.

3.5 *Inductance in series in the line* (Fig. 97)

Let θ_1 and θ_2 be the propagation times across AC and DB, respectively, in Fig. 97. Given the states C(t) and D(t) of points C and D at time t, we wish to construct the states C($t + \Delta t$) and D($t + \Delta t$).

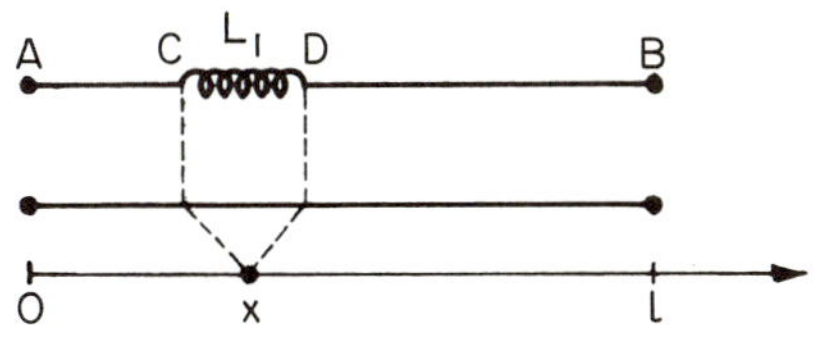

Fig. 97. Line with series lumped inductance.

The average voltage across the inductance is approximately

$$\frac{[v_C(t) - v_D(t)] + [v_C(t + \Delta t) - v_D(t + \Delta t)]}{2} = L_1 \frac{\Delta i}{\Delta t}$$

An observer O_1 leaving A at time $t + \Delta t - \theta_1$ and arriving at C at time $t + \Delta t$ follows the straight line characteristic δ in the (v, i) plane of Fig. 98. An observer O_2 leaving B at $t + \Delta t - \theta_2$ arrives at D at $t + \Delta t$, following the characteristic Δ.

The construction of Fig. 98 proceeds as follows. A line parallel to the i axis is drawn through the point D(t), and a line making an angle $-\alpha$ with

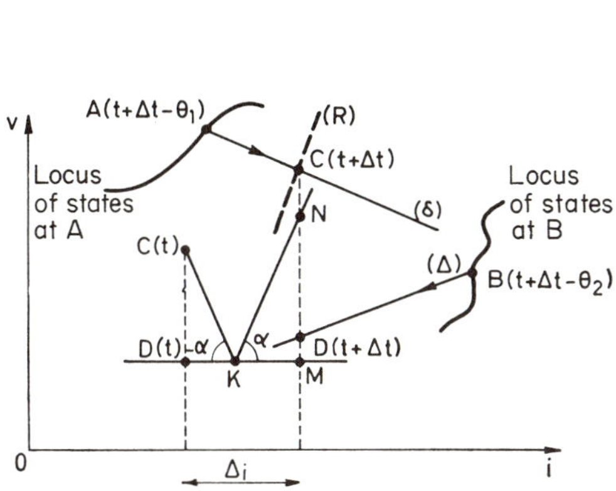

Fig. 98. Portion of state diagram for circuit of Fig. 97.

the i axis is drawn through C(t), where α is given by $\tan \alpha = L_1/(\Delta t/2)$. These two lines intersect at some point K. Then through K a line making an angle $+\alpha$ with the i axis is drawn (the line KN). Using the line D(t)K as reference, or zero ordinate, the ordinates of the two lines KN and Δ are added, to produce the dashed line R. The point C($t + \Delta t$) is then the intersection of the line R and the characteristic δ, and the point D($t + \Delta t$) is the intersection of the characteristic Δ and a vertical through C($t + \Delta t$).

This construction is justified by the following calculation:

$$v_C(t + \Delta t) - v_D(t + \Delta t) = \mathrm{MN}; \qquad \tan \alpha = \frac{2L_1}{\Delta t}$$

$$\tan \alpha = \frac{v_C(t) - v_D(t)}{\mathrm{D}(t)\mathrm{K}} = \frac{\mathrm{MN}}{\mathrm{KM}} = \frac{v_C(t + \Delta t) - v_D(t + \Delta t)}{\mathrm{KM}}$$

$$= \frac{v_C(t) - v_D(t) + v_C(t + \Delta t) - v_D(t + \Delta t)}{\mathrm{D}(t)\mathrm{M}} = \frac{2L_1}{\Delta t}$$

or further

$$\frac{v_C(t) - v_D(t) + v_C(t + \Delta t) - v_D(t + \Delta t)}{2} = \frac{L_1}{\Delta t} \Delta i$$

This is correct to the first order in Δt, since both sides are an approximation to the voltage across the inductance.

It should be remarked that this construction is laborious, and becomes unacceptably so if θ_1 and θ_2 are not in the ratio of small integers.

3.6 *Line loaded at its end by a capacitor* (Fig. 99)

The current through a capacitance C_1 is

$$i = C_1 \frac{dv}{dt}$$

which can be approximated by

$$i = C_1 \frac{\Delta v}{\Delta t}$$

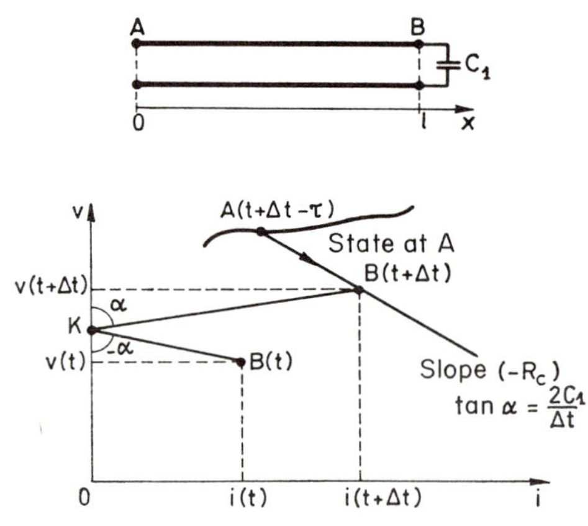

Fig. 99. Line with capacitive load, and state diagram.

Again, this approximation is the better the smaller is Δt, but the graphical construction becomes more difficult for small Δt. Let us assume again that $\Delta t = \tau/n$, where n is an integer.

Rather than repeating the above reasoning, it is only necessary to note the duality:

$$\text{Inductance} \leftrightarrow \text{Capacitance}$$

$$\text{Current} \leftrightarrow \text{Voltage}$$

$$\text{Voltage} \leftrightarrow \text{Current}$$

The results of Section 3.4 can be transcribed using these, to yield the construction of Fig. 99.

3.7 *Capacitor in series in the line* (Fig. 100)

In this case the duality noted in Section 3.6 allows the result of Section 3.5 to be transcribed to yield the construction of Fig. 100.

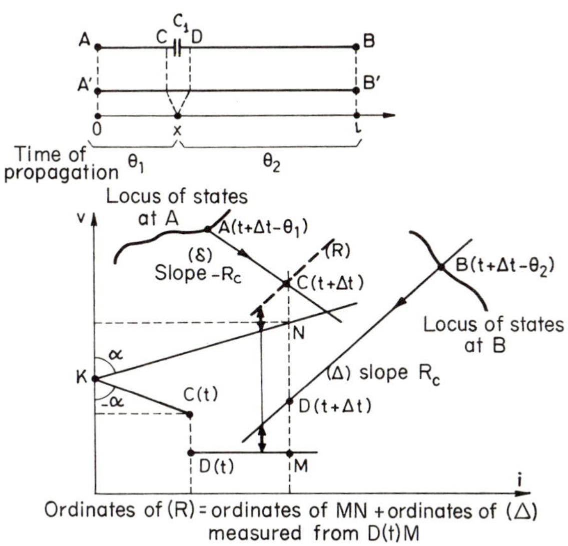

Fig. 100. Line with lumped capacitance in series, and state diagram.

By now it is clear that the method of Bergeron is quite laborious if any reactive elements are present in the circuit. In such cases the method is usually not used. The main applications are in cases with linear or nonlinear resistive elements.

4. Input Signal $e(t)$ Different from a Step Function

Bergeron's method lends itself well to problems in which the input signal is some time function more complex than a simple step.

1. *Principle*

Frequently an actual excitation signal is not well approximated by a step function. Often the input $e(t)$ is a voltage or current ramp function. The method will be explained for the case of a linear voltage ramp. The time axis is assumed divided into equal intervals of length $\Delta t = \tau/n$ or $k\tau$, n and k being integers and τ the propagation time through the line.

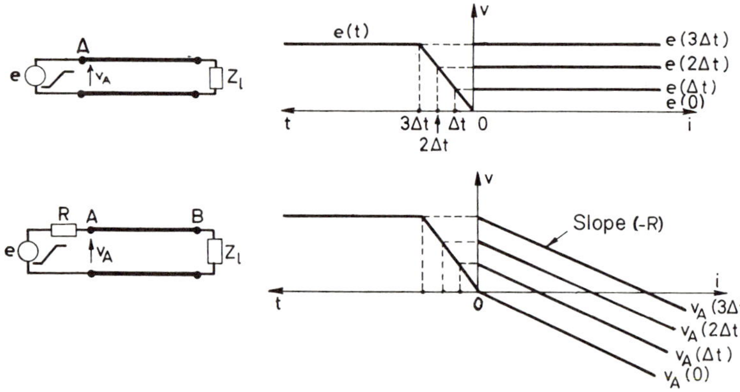

Fig. 101. Time-variable input state diagrams for nonstep sources.

Two graphs are then drawn side by side, having axes (v, i) and (v, t) or (v, i) and (i, t) in the case of current excitation. Two examples are shown in Fig. 101. These allow the state at the input to the line to be determined at the times 0, Δt, $2\Delta t$, ... Observers are then sent on their journeys, starting from A at times 0, Δt, $2\Delta t$, ...

2. *First example* (Figs. 102 and 103)

Consider a line terminated in a resistance $R_l = R_c/2$ and excited by a voltage ramp of duration 4τ, produced by a generator with zero internal

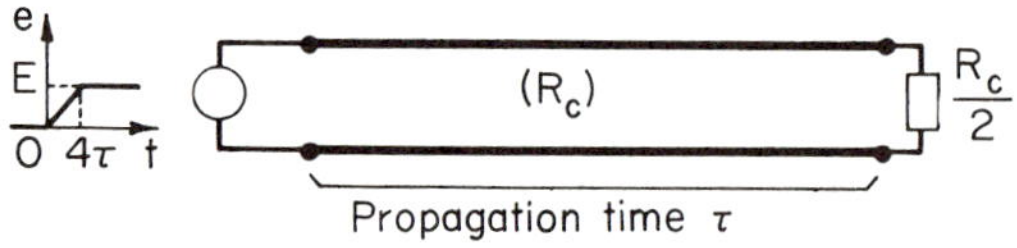

Fig. 102. Line with ramp excitation.

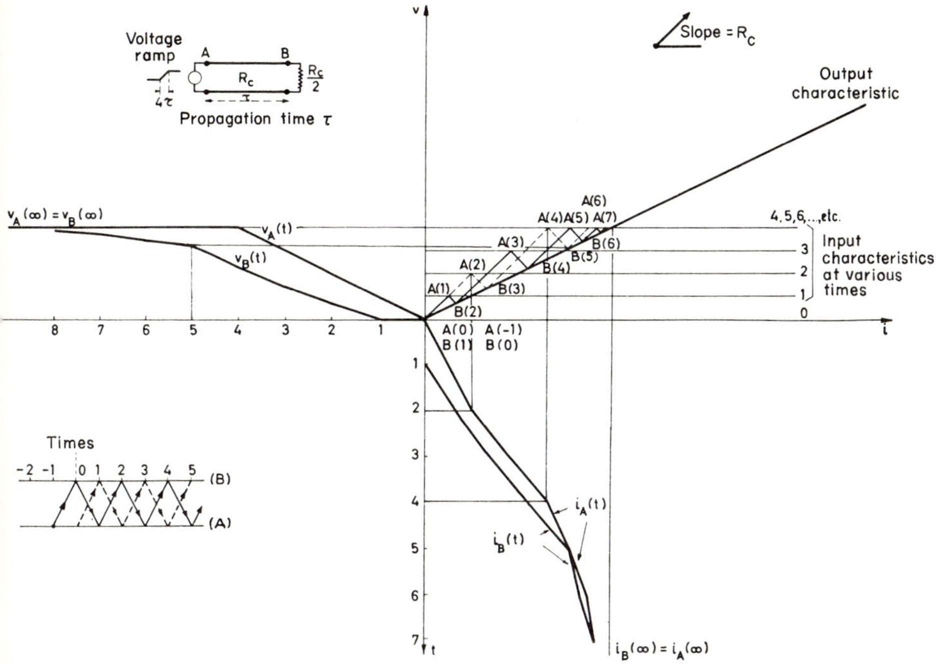

Fig. 103. Time chart and Bergeron diagram for circuit of Fig. 102.

resistance, as in Fig. 102. Two observers suffice, leaving A at times $-\tau$ and 0. Figure 103 has been drawn for the case $\tau = 1$.

3. *Second example* (Figs. 104 and 105)

Consider a line terminated in $R_l = R_c/2$ and excited by a current ramp of duration 4τ produced by a generator with infinite internal resistance, as in Fig. 104. Again two observers suffice, leaving A at times $-\tau$ and 0. The result is shown in Fig. 105.

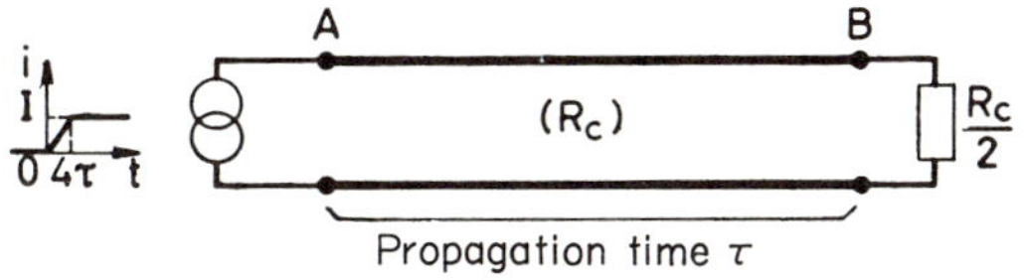

Fig. 104. Line excited by current ramp.

4. *Third example* (Figs. 106 and 107)

Consider the circuit of Fig. 106. Three observers are necessary, leaving A at times $-\tau$, $-\tau/2$, and 0. The result is shown in Fig. 107.

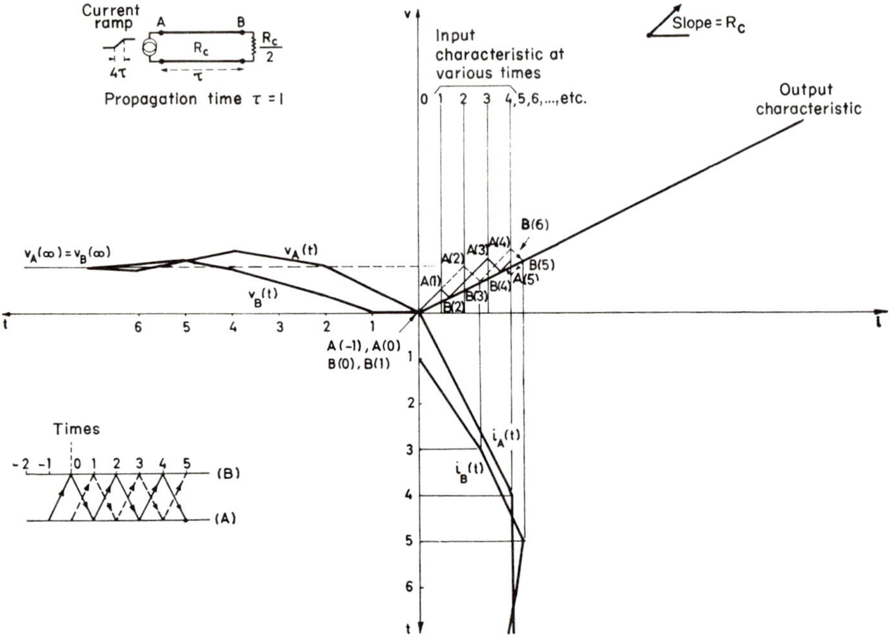

Fig. 105. Time chart and Bergeron diagram for circuit of Fig. 104.

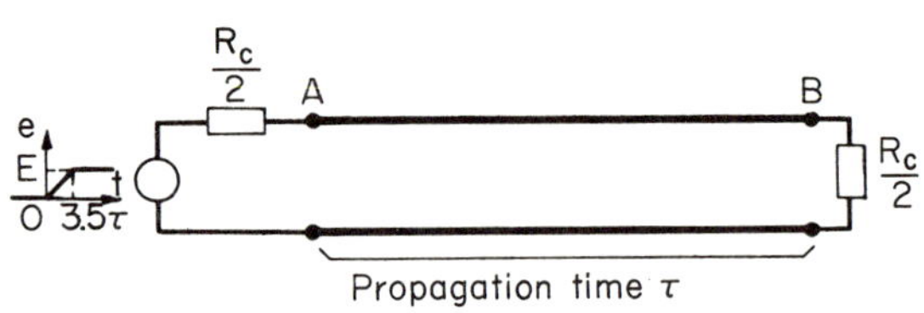

Fig. 106. Line driven by non-ideal ramp voltage source.

Exercises

The solutions to exercises marked* will be found in Chapter VI.

*Exercise 1. Consider a lossless line, open-circuited at the output, and previously charged to a voltage E by a battery, as in Fig. 108. At time $t = 0$, the switch is thrown, grounding the input. Find the output voltage and input current as a function of time.

*Exercise 2. A generator with internal resistance $R < R_c$, and having an open-circuit voltage which is a step $+E$, is used to excite a lossless line

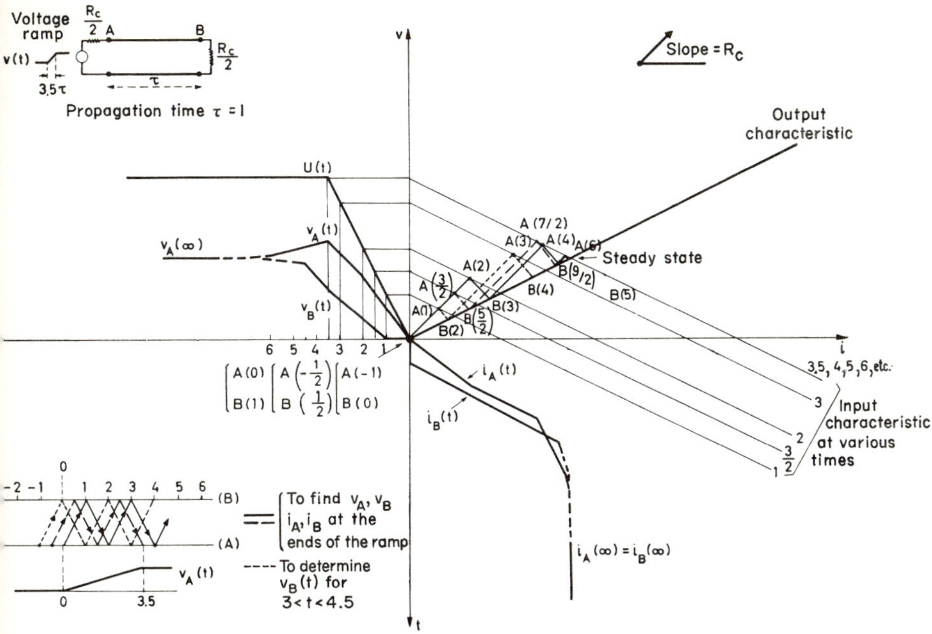

Fig. 107. Time chart and Bergeron diagram for circuit of Fig. 106.

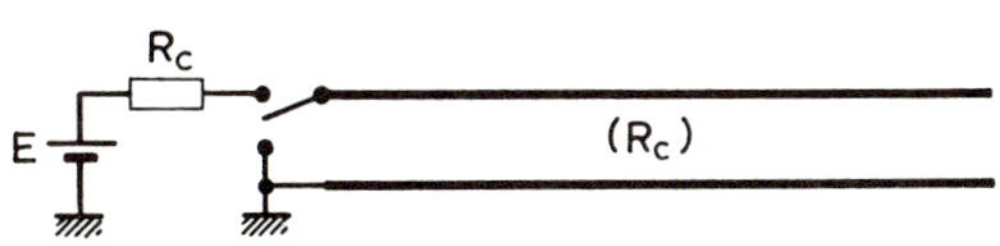

Fig. 108. Circuit for Exercise 1.

Fig. 109. Circuit for Exercise 2.

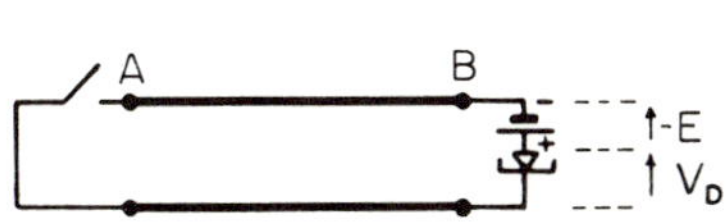

Fig. 110. Circuit for Exercise 6.

with characteristic resistance R_c. The output of the line is short-circuited (Fig. 109). Find the voltage and current at the input and output of the line.

*EXERCISE 3. Solve Exercise 2 of Chapter II, using Bergeron's method.

EXERCISE 4. Solve Exercises 2 and 3 at the end of Chapter IV using Bergeron's method.

*EXERCISE 5. Consider a lossless line with characteristic resistance R_c, terminated in a resistance $R_c/4$, and excited by a generator having internal

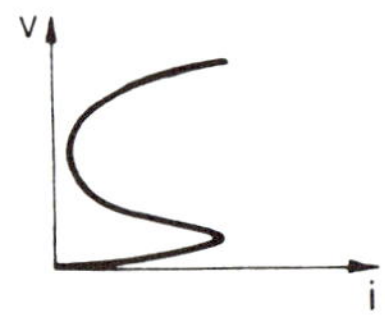

Fig. 111. Volt-ampere characteristic of tunnel diode of Fig. 110.

impedance $R_c/4$ and an open-circuit voltage which is a step $+E$ occurring at time 0. Find the voltage and current at the input and output of the line.

*EXERCISE 6. Consider a lossless line charged at its output end by a tunnel diode biased by a voltage source E. At $t = 0$, a switch is closed which short-circuits the input to the line (Fig. 110). Show that, depending on the value of E, two types of behavior can be obtained: (1) a stable steady state, established after a short transient; (2) a steady oscillation, established after a short transient. The $v(i)$ characteristic of the tunnel diode is shown in Fig. 111.

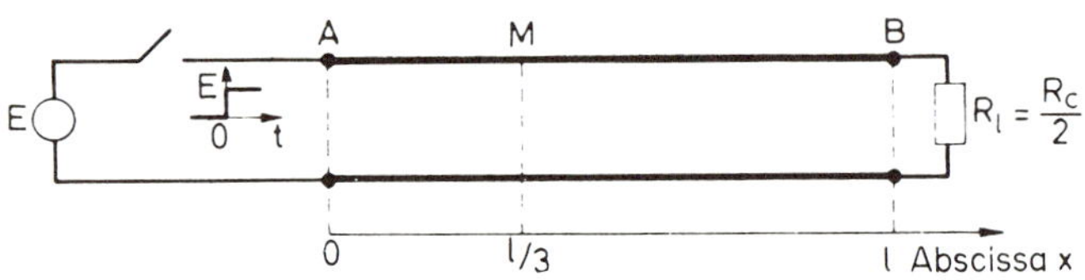

Fig. 112. Circuit for Exercise 7.

*EXERCISE 7. Consider the circuit of Fig. 112. Find the voltage and current at point M.

CHAPTER IV

Line Transients Using Operational Calculus

1. Generalities

1.1 *Introduction*

This chapter concerns the transient behavior of transmission lines, studied using the operational calculus. The reader is assumed to be familiar with the basic Laplace transform techniques, as available in any of a multitude of texts.

Equations (I.29) and (I.30) established in Chapter I are the starting point for the developments of the present chapter. They incorporate all the properties of lines which we have discussed in the preceding chapters, such as propagation and reflection of waves. In addition, they yield the characteristics of the beginning of the transient wave in lossy lines and, in particular, show the effect of skin-effect losses on the step response of the line.

The problem of determining the transient behavior of real lossy lines is approached obliquely, through the study of lossless lines. The latter leads to simple explanations of the various physical phenomena, using the ideas of wave propagation and reflection, and of matching. These studies provide a form of solution which is sufficiently general as to extend, with modifications, to the case of lossy lines. This solution is embodied in Eqs. (IV.5) and (IV.6).

The most important difference between the solution for the lossless line and the behavior actually observed in a real line is the deterioration of the wave amplitudes in the latter case. This deterioration can be accounted for by introducing the skin-effect losses.

1.2 *Summary of the basic relations. General method for study of the transient regime*

The following relations were given in Section 5.4 of Chapter I. They are rewritten here, along with the sign convention used (Fig. 113).

$$V(x, p) = E(p)\frac{Z_c}{Z_0 + Z_c}\,[e^{-\gamma x} + \Gamma_l e^{-\gamma(2l-x)}]\sum_{n=0}^{\infty}\Gamma_0^n\Gamma_l^n e^{-2n\gamma l} \tag{I.29}$$

$$I(x, p) = E(p)\frac{1}{Z_0 + Z_c}\,[e^{-\gamma x} - \Gamma_l e^{-\gamma(2l-x)}]\sum_{n=0}^{\infty}\Gamma_0^n\Gamma_l^n e^{-2n\gamma l} \tag{I.30}$$

with

$$\gamma = [(R + Lp)(G + Cp)]^{1/2} \tag{I.16}$$

$$Z_c = \left(\frac{R + Lp}{G + Cp}\right)^{1/2} \tag{I.20}$$

$$\Gamma_0 = \frac{Z_0 - Z_c}{Z_0 + Z_c} \tag{I.23}$$

$$\Gamma_l = \frac{Z_l - Z_c}{Z_l + Z_c} \tag{I.25}$$

In general Z_0, Z_l, Γ_0, and Γ_l are functions of the Laplace transform variable p.

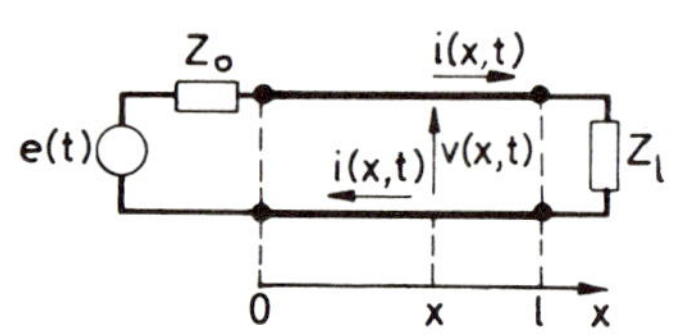

Fig. 113. Current and voltage sign conventions.

Having decided to use the operational calculus for solving some transmission line problem, the procedure is as follows. First apply the above formulas to the particular problem at hand, then determine the inverse transforms of $V(x, p)$ and $I(x, p)$ to obtain the time behavior $v(x, t)$ and $i(x, t)$. In general the expressions to which one is led are quite complicated. We will show, however, that in many cases various simplifications are possible, so that the inverse transforms follow immediately.

1.3 *Theoretical developments related to the characteristic impedance* Z_c

(a) Case of a Line Terminated in a Lumped Impedance Equal to the Characteristic Impedance Z_c. Applying the above formulas with $Z_l = Z_c$ yields $\Gamma_l = 0$, so that

$$V(x, p) = E(p)\frac{Z_c}{Z_0 + Z_c} e^{-\gamma x}; \qquad I(x, p) = E(p)\frac{1}{Z_0 + Z_c} e^{-\gamma x}$$

from which

$$\frac{V(x, p)}{I(x, p)} = Z_c$$

Thus we have the following result. At each instant t, at each point x of a line terminated in its characteristic impedance, the ratio of the voltage across the line to the current through the line is equal to the characteristic impedance. This result is of theoretical interest only, except in a few special cases, such as lossless lines and lines at the beginning of the transient regime, since it is not possible to realize a lumped-parameter circuit with impedance

$$Z_c = \left(\frac{R + Lp}{G + Cp}\right)^{1/2}$$

(b) The Semi-Infinite Line ($l = \infty$) (Fig. 114). For this case we obtain

$$V(x, p) = E(p)\frac{Z_c}{Z_0 + Z_c} e^{-\gamma x}$$

$$I(x, p) = E(p)\frac{1}{Z_0 + Z_c} e^{-\gamma x}$$

These are identical to the relations obtained above for a line of finite length terminated in its characteristic impedance. Thus we obtain the result that a line of infinite length behaves like a line of finite length terminated in its characteristic impedance.

Fig. 114. A semi-infinite line.

Remark. Here again this result is theoretical, since in any real line the end is not at infinity. It is however of practical use if the wave reflected at the load decays sufficiently in its passage there and back along the line as to be negligible when it arrives at the input (see Section 3.5).

2. Lossless Lines. General Properties

2.1 *Generalities*

(a) Definition. A line for which $R = 0$ and $G = 0$ is said to be *lossless.*

(b) Characteristic Resistance. The impedance

$$Z_c = \left(\frac{R + Lp}{G + Cp}\right)^{1/2}$$

reduces in this case to a pure resistance, called the *characteristic resistance*:

$$R_c = (L/C)^{1/2} \tag{IV.1}$$

In the case of a lossless line, the two properties discussed in Section 1.3 can thus be stated as follows. At each instant and at each point of a lossless line, either terminated in its characteristic resistance R_c or of infinite length, the ratio of the voltage across the line to the current through the line is R_c.

(c) Propagation Constant. The general expression

$$\gamma = [(R + Lp)(G + Cp)]^{1/2}$$

becomes for the lossless line

$$\gamma = p(LC)^{1/2} = p\delta \tag{IV.2}$$

where we define $\delta = (LC)^{1/2}$. γ is called the *propagation constant*, and has the dimensions of an inverse velocity.

2.2 *Equations for the lossless line*

Taking account of the simplifications and notation introduced in Section 2.1, the fundamental relations (I.27)–(I.30) become for the lossless line

$$V(x, p) = E(p)\frac{R_c}{Z_0 + R_c}\,\frac{e^{-p\delta x} + \Gamma_l e^{-p\delta(2l-x)}}{1 - \Gamma_0\Gamma_l e^{-2p\delta l}} \tag{IV.3}$$

$$I(x, p) = \frac{E(p)}{Z_0 + R_c}\,\frac{e^{-p\delta x} - \Gamma_l e^{-p\delta(2l-x)}}{1 - \Gamma_0\Gamma_l e^{-2p\delta l}} \tag{IV.4}$$

$$V(x, p) = E(p)\frac{R_c}{Z_0 + R_c}\,[e^{-p\delta x} + \Gamma_l e^{-p\delta(2l-x)}]\sum_{n=0}^{\infty} \Gamma_0{}^n\Gamma_l{}^n e^{-2np\delta l} \tag{IV.5}$$

$$I(x, p) = E(p)\frac{1}{Z_0 + R_c}\,[e^{-p\delta x} - \Gamma_l e^{-p\delta(2l-x)}]\sum_{n=0}^{\infty}\Gamma_0^n \Gamma_l^n e^{-2np\delta l} \qquad \text{(IV.6)}$$

$$R_c = (L/C)^{1/2}$$

$$\Gamma_0 = \frac{Z_0 - R_c}{Z_0 + R_c}$$

$$\delta = (LC)^{1/2}$$

$$\Gamma_l = \frac{Z_l - R_c}{Z_l + R_c}$$

2.3 *Physical interpretation of the propagation constant* δ

(1) Consider a lossless line with a matched load ($Z_l = R_c$). Then $\Gamma_l = 0$, and the general equations become

$$V(x, p) = E(p)\frac{R_c}{Z_0 + R_c}\,e^{-p\delta x}$$

$$I(x, p) = E(p)\frac{1}{Z_0 + R_c}\,e^{-p\delta x} = \frac{V(x, p)}{R_c}$$

Hence $i(x, t) = v(x, t)/R_c$, and it suffices to calculate $V(x, p)$.

The term $e^{-p\delta x}$ in $V(x, p)$ corresponds to a delay δx. The transform of $v(0, t)$, the voltage at the input to the line, is

$$E(p)\frac{R_c}{Z_0 + R_c}$$

hence

$$v(0, t) = e(t)\frac{R_c}{Z_0 + R_c}$$

Knowing the voltage at the input to the line, the voltage at any abscissa x along the line follows from

$$v(x, t) = v(0, t - \delta x)\,\Upsilon(t - \delta x)$$

(Fig. 115). Here $\Upsilon(t - \delta x)$ is the unit step function beginning at time δx.

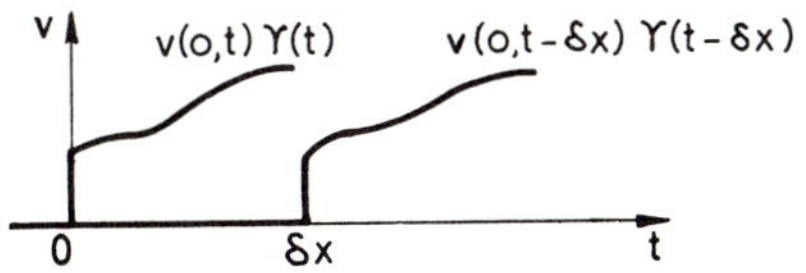

Fig. 115. Voltage waveforms along a matched lossless line.

Thus the voltage and current at time t and abscissa x along the line are the same as the voltage and current at the input $x = 0$ at time $t - \delta x$. Hence a disturbance propagates without distortion from the input $x = 0$ to an arbitrary abscissa x in time δx, so that $1/\delta$ is the velocity of propagation along the line:

$$u = \frac{1}{\delta} = \frac{1}{(\mathrm{LC})^{1/2}} \tag{IV.7}$$

(2) Propagation velocity. It can be shown† that if the line is surrounded by a dielectric medium with relative dielectric constant ε_r, disturbances propagate along the line with the same velocity as electromagnetic waves in the dielectric medium, which is to say

$$u = \frac{c}{\varepsilon^{1/2}}$$

from which

$$\delta = \frac{1}{u} = \frac{\varepsilon^{1/2}}{c} \tag{IV.8}$$

Here c is the speed of light in vacuum.

2.4 *Expressions for* L *and* C *as functions of* R_c *and* δ

Recall that L and C are the self-inductance per unit length and capacitance per unit length of the line. We have

$$\mathrm{R}_c = (\mathrm{L/C})^{1/2}, \qquad \delta = (\mathrm{LC})^{1/2}$$

$$\mathrm{R}_c^{\,2} = \mathrm{L/C}, \qquad \delta^2 = \mathrm{LC}$$

from which

$$\mathrm{L} = \mathrm{R}_c \delta \tag{IV.9}$$

$$\mathrm{C} = \frac{\delta}{\mathrm{R}_c} \tag{IV.10}$$

For a line of length l, the delay is $\tau = \delta l$, the total inductance is $\mathscr{L} = \mathrm{L}l = \mathrm{R}_c\tau$, and the total capacitance is $\mathscr{C} = \mathrm{C}l = \tau/\mathrm{R}_c$.

Expressions (IV.9) and (IV.10) are of frequent use.

From (IV.10) it can be seen that since for a coaxial cable R_c depends on the diameters d_1 and d_2 (Fig. 116), so also does the capacitance per unit

† G. Bruhat and G. Goudet, "Électricité." Masson, Paris, 1963.

length C. In particular, from the formulas of Section 6.3 of Chapter I, we have

$$C = \frac{2\pi\varepsilon_0\varepsilon_r}{\text{Log}\,\dfrac{d_2}{d_1}}; \qquad R_c = \frac{1}{2\pi}\left(\text{Log}\,\frac{d_2}{d_1}\right)(\mu_0/\varepsilon_r\varepsilon_0)^{1/2}$$

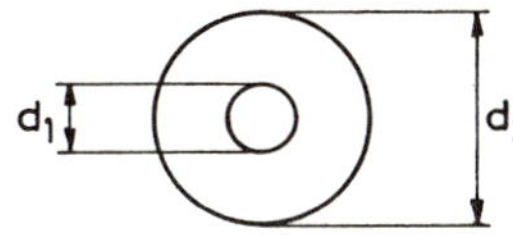

Fig. 116. Geometry of a coaxial cable.

from which

$$R_c C = \frac{\varepsilon^{1/2}}{3 \times 10^8}$$

In the same manner we obtain from $L = CR_c^2$

$$L = \frac{\sqrt{\varepsilon_r}}{3 \times 10^8}\ R_c$$

For cables with Teflon insulation ($\varepsilon_r = 2.25$), this becomes

$$C = \frac{5 \times 10^{-9}}{R_c} = \frac{5000}{R_c}\,10^{-12} \quad (\text{F}), \qquad L = 5 \times 10^{-9}\, R_c \quad (\text{H})$$

Examples. 50-Ω coaxial cable. The dielectrics used in these cables have an ε_r between 2 and 2.5, so we will use as an average $\varepsilon_r = 2.25$. Then

$$\delta = \frac{\varepsilon^{1/2}}{c} = \frac{(2.25)^{1/2}}{3 \times 10^8} = 5 \times 10^{-9} \quad \text{sec/meter}$$

$$\delta = 5 \quad \text{nsec/meter}$$

$$C = \frac{\delta}{R_c} = \frac{5 \times 10^{-9}}{50} = 100 \quad \text{pF/meter}$$

$$L = \delta R_c = 50 \times 5 \times 10^{-9} = 250 \quad \text{nH/meter}$$

or

$$L = 50 \quad \text{nH/nsec delay}$$

Note that these values are independent of the dimensions of the cable, and, in particular, of the length.

2.5 *Principal results*

(a) LINE WITH MATCHED LOAD $Z_l = R_c$ (Fig. 117). It was shown in Section 1.3 of Chapter II and is discussed in Exercise 1 of the present Chapter that the voltage (and the current) at an arbitrary point along the line reproduces the voltage (and current) at the input, with a delay equal to the time of propagation from the input to the point of interest. When the disturbance arrives at the end $x = l$, it is not reflected.

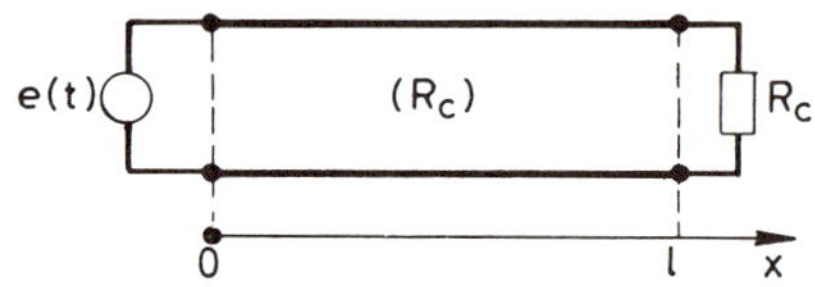

Fig. 117. Line with ideal source and matched load.

(b) LINE OPEN AT THE END (Fig. 118). It can be shown (Exercise 2) that at the output the voltage is reflected without change of sign, and the current with change of sign.

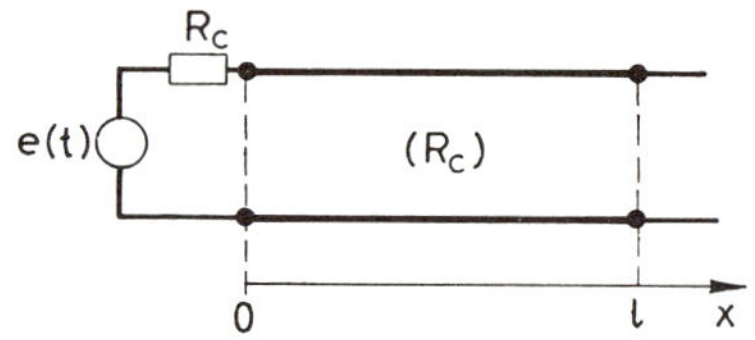

Fig. 118. Open line with matched source.

(c) LINE SHORT-CIRCUITED AT THE OUTPUT (Fig. 119). It can be shown (Exercise 3) that at the output the voltage is reflected with change of sign and the current without change of sign.

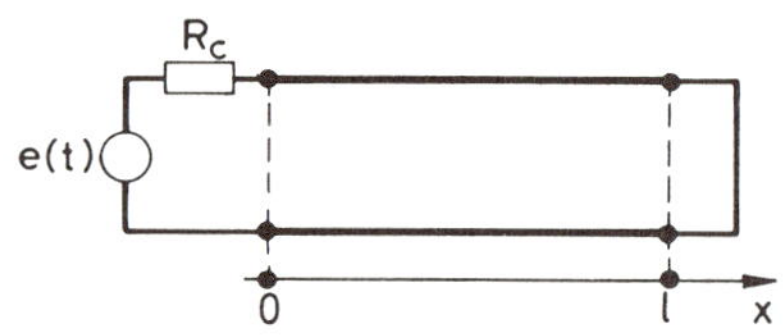

Fig. 119. Shorted line with matched source.

(d) INPUT IMPEDANCE DURING THE TIME INTERVAL $0 \leq t \leq 2\tau$. Consider a lossless line which is matched neither at its input nor at its output.

The transform of the voltage at the input to the line is

$$V(0, p) = E(p) \frac{R_c}{Z_0 + R_c} (1 + \Gamma_l e^{-2\delta l p}) \sum_{n=0}^{\infty} \Gamma_0^n \Gamma_l^n e^{-2np\delta l}$$

Using $\delta l = \tau$, this becomes

$$V(0, p) = E(p) \frac{R_c}{Z_0 + R_c} \left[1 + \Gamma_l e^{-2\tau p} + (1 + \Gamma_l e^{-2\tau p}) \sum_{n=1}^{\infty} \Gamma_0^n \Gamma_l^n e^{-2np\tau}\right]$$

Only the first term in this expression,

$$E(p) \frac{R_c}{Z_0 + R_c}$$

contributes to the time response on the interval $0 \leq t \leq 2\tau$, that is, to the response before the first excitation has had time to make a round trip to the end of the line and back. Thus for $0 \leq t \leq 2\tau$, the input voltage is

$$V(0, p) = E(p) \frac{R_c}{Z_0 + R_c}$$

which shows that during this time the line behaves like a resistance R_c. This is the case regardless of the length of the line and the load impedance Z_l.

Thus again, a lossless line, mismatched at the output, behaves at the input like a resistance equal to the characteristic resistance R_c, during the time necessary for the disturbance applied at the input to propagate to the output and return.

2.6 *Examples*

1. *Line matched at the input and loaded with a reactive load*

Consider the circuit of Fig. 120. The generator furnishes a voltage step $E\Upsilon(t)$ beginning at $t = 0$. The line is loaded by a capacitance C_1 in series with a resistance R_1. We wish to find $v(l, t)$ and $i(l, t)$.

For this case,

$$Z_l = \frac{1}{C_1 p} + R_1; \qquad \Gamma_0 = 0$$

$$\Gamma_l = \frac{\dfrac{1}{C_1 p} + R_1 - R_c}{\dfrac{1}{C_1 p} + R_1 + R_c} = 1 - \frac{2R_c C_1 p}{1 + C_1 p(R_1 + R_c)}$$

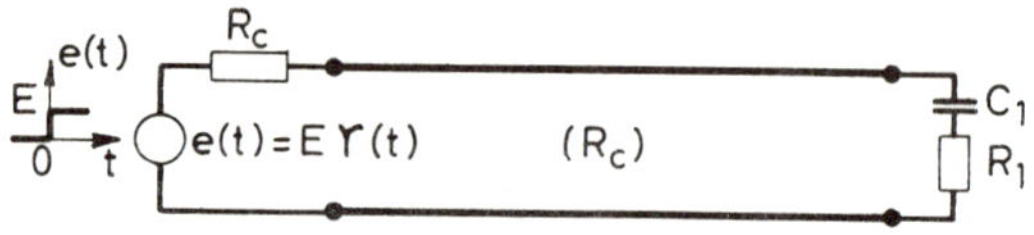

Fig. 120. Line with capacitive load and matched source.

Relations (IV.3) and (IV.4) become

$$V(l, p) = \frac{E}{p}\,\frac{1}{2}\,(1 + \Gamma_l)e^{-\tau p} \qquad \text{with} \quad \tau = \delta l$$

$$I(l, p) = \frac{E}{p}\,\frac{1}{2R_c}\,(1 - \Gamma_l)e^{-\tau p}$$

Substituting for Γ_l yields

$$V(l, p) = \frac{E}{p}\left(1 - \frac{R_c C_1 p}{1 + C_1 p(R_1 + R_c)}\right)e^{-\tau p}$$

$$I(l, p) = E\,\frac{C_1}{1 + C_1 p(R_1 + R_c)}\,e^{-\tau p}$$

or

$$V(l, p) = E\left[\frac{1}{p} - \frac{R_c}{R_1 + R_c}\,\frac{1}{p + \dfrac{1}{C_1(R_1 + R_c)}}\right]e^{-\tau p}$$

$$I(l, p) = \frac{E}{R_1 + R_c}\,\frac{1}{\left[p + \dfrac{1}{C_1(R_1 + R_c)}\right]}\,e^{-\tau p}$$

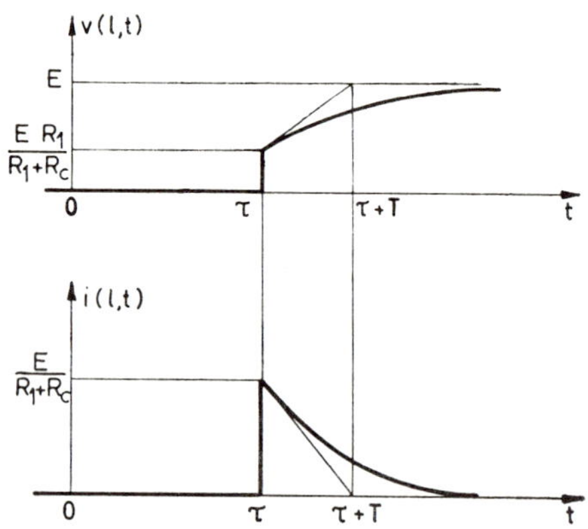

Fig. 121. Load voltage and current for circuit of Fig. 120.

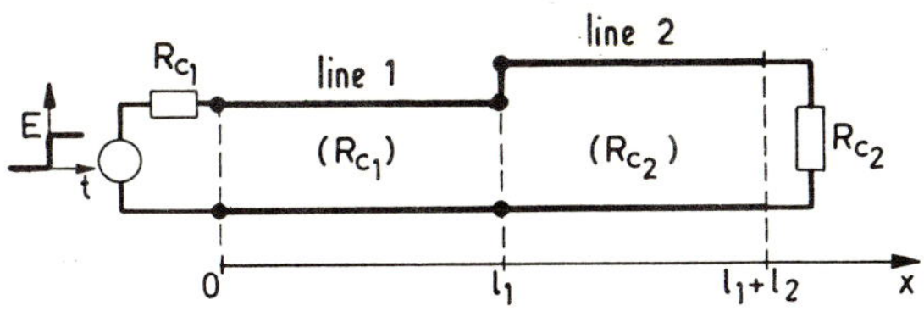

Fig. 122. Line with discontinuous characteristic resistance.

Defining

$$\frac{1}{C_1(R_1 + R_c)} = \frac{1}{T}$$

yields finally

$$v(l, t) = E\left[1 - \frac{R_c}{R_1 + R_c} e^{-(t-\tau)/T}\right]\Upsilon(t - \tau)$$

$$i(l, t) = \frac{E}{R_1 + R_c} e^{-(t-\tau)/T}\Upsilon(t - \tau)$$

These are shown in Fig. 121.

2. *Change of characteristic impedance*

Consider a lossless line consisting of two sections with different characteristic impedances R_{c1} and R_{c2}, as shown in Fig. 122. Line 1 is of length l_1, and is matched at the input. Line 2, of length l_2, is matched at the output. The generator furnishes a voltage step $E\Upsilon(t)$.

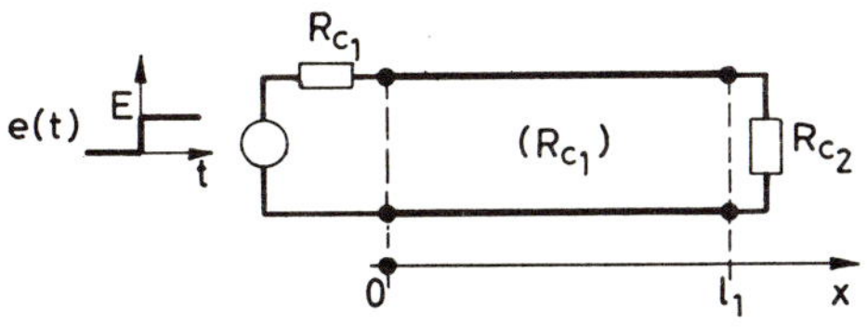

Fig. 123. Circuit equivalent to that of Fig. 122, so far as line 1 is concerned.

Let us first examine the situation in line 1. Seen from its input, line 2 is equivalent to a resistance R_{c2}. Hence so far as line 1 is concerned, the circuit of Fig. 122 can be replaced by that of Fig. 123. Applying relations (IV.3) and (IV.4) to this latter circuit, for which

$$\Gamma_0 = 0; \qquad \Gamma_{l_1} = \frac{R_{c2} - R_{c1}}{R_{c2} + R_{c1}}; \qquad \Gamma_{l_1+l_2} = 0$$

and defining

$$\delta l_1 = \tau_1 \qquad \delta l_2 = \tau_2 \qquad \Gamma_{l_1} = \Gamma$$

yields at the input to the line

$$V(0, p) = \frac{E}{2p}(1 + \Gamma e^{-2\tau_1 p}) \qquad \Rightarrow v(0, t) = \frac{E}{2}[\Upsilon(t) + \Gamma\Upsilon(t - 2\tau_1)]$$

$$I(0, p) = \frac{E}{2pR_{c1}}(1 - \Gamma e^{-2\tau_1 p}) \Rightarrow i(0, t) = \frac{E}{2R_{c1}}[\Upsilon(t) - \Gamma\Upsilon(t - 2\tau_1)] \qquad \text{(IV.11)}$$

and at the output,

$$V(l_1, p) = \frac{E}{2p}(1 + \Gamma)e^{-\tau_1 p} \qquad \Rightarrow v(l_1, t) = \frac{E}{2}(1 + \Gamma)\Upsilon(t - \tau_1)$$

$$I(l_1, p) = \frac{E}{2pR_{c1}}(1 - \Gamma)e^{-\tau_1 p} \Rightarrow i(l_1, t) = \frac{E(1 - \Gamma)\Upsilon(t - \tau_1)}{2R_{c1}} \qquad \text{(IV.12)}$$

Relations (IV.11) and (IV.12) describe affairs at the input and output of line 1. They show that at $t = 0$ a voltage step appears at the input, which propagates towards the output where it is reflected with reflection coefficient Γ. The reflected step returns to the input, where it is absorbed.

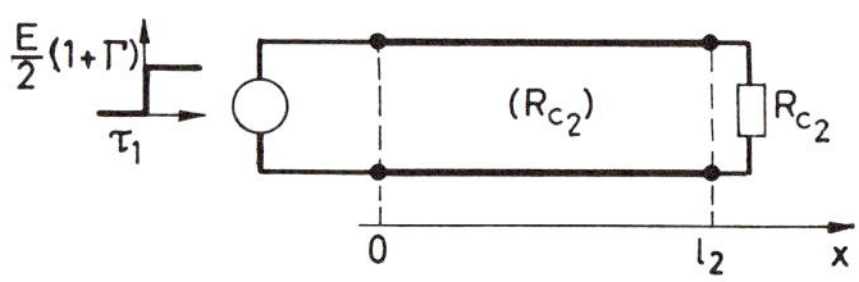

Fig. 124. Circuit equivalent to that of Fig. 122, so far as line 2 is concerned.

Let us now consider line 2. At $t = \tau_2$, according to (IV.12) there appears at the input to line 2 a voltage step $E(1 + \Gamma)/2$, which causes a current step $E(1 + \Gamma)/(2R_{c2})$ to enter line 2. So far as line 2 is concerned, the circuit of Fig. 122 can be replaced with that of Fig. 124. The step $E(1 + \Gamma)/2$ propagates towards the output of line 2, where it arrives at time $(\tau_1 + \tau_2)$ and is absorbed by the resistance R_{c2}.

The final results obtained are shown in Fig. 125.

Application. Knowing the characteristic impedance R_{c1} of line 1, by observing the input voltage $v(0, t)$ to line 1 on an oscilloscope, it is possible to determine the characteristic impedance R_{c2} of line 2. From observation of $v(0, t)$, it is possible to determine

$$\Gamma = \frac{v(0, \infty) - E/2}{E/2}$$

But

$$\Gamma = \frac{R_{c2} - R_{c1}}{R_{c2} + R_{c1}}$$

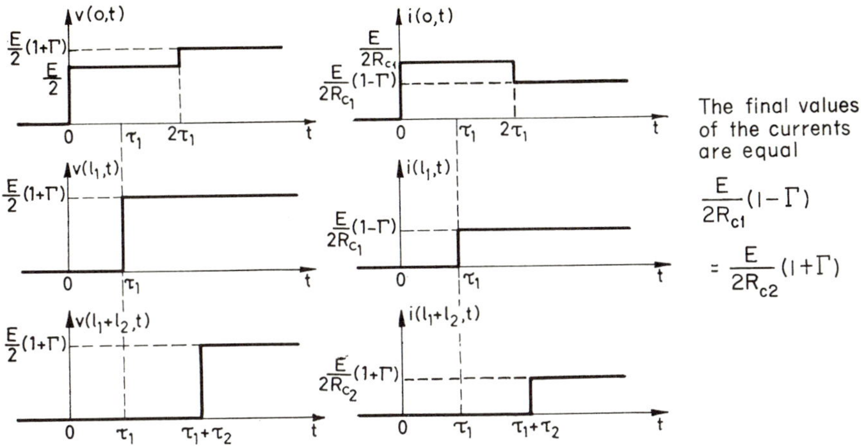

Fig. 125. Voltage and current waveforms in circuit of Fig. 122.

so that

$$R_{c2} = R_{c1} \frac{1 + \Gamma}{1 - \Gamma} \tag{IV.13}$$

Remark. The relations governing the propagation of waves across the junction between the two lines can also be obtained by requiring that energy be conserved across the junction (see Chapter II, Section 3.6).

2.7 *Multiple reflections in a lossless line*

1. *The state at an arbitrary point*

In all the examples considered so far, reflections occurred only at the output, the input being always matched. It has always been the case that $\Gamma_0 = 0$, so that the denominator of relations (IV.3) and (IV.4) reduced to 1. Multiple reflections are said to be present if reflections occur at the input to the line as well as at the output. This is the case if $\Gamma_0 \neq 0$ and $\Gamma_l \neq 0$, which is to say if $Z_0 \neq R_c$ and $Z_l \neq R_c$.

To study these cases of multiple reflections on a lossless line, it is necessary to go back to the basic relations

$$V(x, p) = E(p) \frac{R_c}{Z_0 + R_c} [e^{-p\delta x} + \Gamma_l e^{-p\delta(2l-x)}] \sum_{n=0}^{\infty} \Gamma_0^n \Gamma_l^n e^{-2np\delta l} \tag{IV.5}$$

$$I(x, p) = E(p) \frac{1}{Z_0 + R_c} [e^{-p\delta x} - \Gamma_l e^{-p\delta(2l-x)}] \sum_{n=0}^{\infty} \Gamma_0^n \Gamma_l^n e^{-2np\delta l} \tag{IV.6}$$

When the impedances Z_0 and Z_l are general, the reflection coefficients Γ_0 and Γ_l are functions of the complex frequency variable p. It is then very difficult to compute the inverse transforms of $V(x, p)$ and $I(x, p)$. In order better to explain the establishment of multiple reflections on a line, we will assume that Z_0 and Z_l are pure resistances R_0 and R_l. Then the reflection coefficients do not involve p:

$$\Gamma_0 = \frac{R_0 - R_c}{R_0 + R_c}; \qquad \Gamma_l = \frac{R_l - R_c}{R_l + R_c}$$

Note that always

$$-1 \leq \Gamma_0 \leq +1$$

$$-1 \leq \Gamma_l \leq +1$$

Let

$$E(p) \frac{R_c}{R_0 + R_c} = V_0(p) \Rightarrow v_0(t)$$

be the input voltage to the line for $0 < t < 2\tau$, and

$$E(p) \frac{1}{R_0 + R_c} = I_0(p) \Rightarrow i_0(t)$$

be the current entering into the line at $t = 0$. Relation (IV.5) then becomes

$$V(x, p) = V_0(p) [e^{-p\delta x} + \Gamma_l e^{-p\delta(2l-x)}] \sum_{n=0}^{\infty} \Gamma_0^n \Gamma_l^n e^{-2np\delta l}$$

or

$$V(x, p) = V_0(p) \sum_{n=0}^{\infty} \Gamma_0^n \Gamma_l^n e^{-\delta(2nl+x)p} + V_0(p) \sum_{n=0}^{\infty} \Gamma_0^n \Gamma_l^{n+1} e^{-\delta[2(n+1)l-x]p}$$

The inverse transform follows at once:

$$v(x, t) = \sum_{n=0}^{\infty} \Gamma_0^n \Gamma_l^n v_0[t - \delta(2nl + x)] \Upsilon[t - \delta(2nl + x)]$$

$$+ \sum_{n=0}^{\infty} \Gamma_0^n \Gamma_l^{n+1} v_0[t - \delta[2(n + 1)l - x]] \Upsilon[t - \delta[2(n + 1)l - x]]$$

This latter expression shows that the voltage at an arbitrary point x along the line (Fig. 126) is the sum of infinitely many voltage waves, each

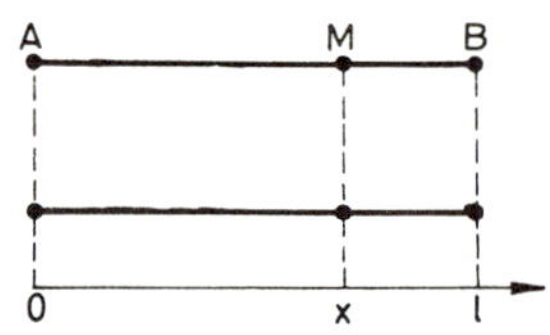

Fig. 126. General point on a line.

of the same waveform as the input voltage $v_0(t)$, but having different amplitudes and appearing at successive instants in time.

Writing the above voltage out in full yields

$$\begin{aligned} v(x, t) = {} & v_0(t - \delta x)\Upsilon(t - \delta x) + \Gamma_l v_0[t - \delta(2l - x)]\Upsilon[t - \delta(2l - x)] \\ & + \Gamma_0\Gamma_l v_0[t - \delta(2l + x)]\Upsilon[t - \delta(2l + x)] \\ & + \Gamma_0\Gamma_l^2 v_0[t - \delta(4l - x)]\Upsilon[t - \delta(4l - x)] + \cdots \end{aligned} \tag{IV.14}$$

The first term in the first line represents the wave $v_0(t)$ arriving without deformation at the point x, at time $t = \delta x$, after having traversed the path Ox. The second term of the first line represents the perturbation $\Gamma_l v_0(t)$ which arrives at x at time $t = \delta(2l - x)$, after traversing $(Ol + lx)$. The reflection process at the end of the line accounts for the amplitude decrease factor Γ_l. The first term of the second line represents the disturbance $\Gamma_0\Gamma_l v_0(t)$ which arrives at x at time $t = \delta(2l + x)$, after traversing the path $Ol + lO + Ox$. Reflection at the load, and then at the input, multiplies the amplitude by Γ_l, then by Γ_0. This process of multiple reflections then continues.

For some given finite time t, the last term of (IV.14) which must be considered is that which is followed by the first step function of value zero.

In relation (IV.14), the waves appearing at the point x are separated into two categories, corresponding to the two columns of the equation. The first class includes waves which have undergone an equal number of reflections at the input and at the output, while the second class consists of waves which have undergone an additional reflection at the output.

Starting from relation (IV.6), and introducing

$$i_0(t) \Leftrightarrow \frac{\mathrm{E}(p)}{\mathrm{R}_0 + \mathrm{R}_c}$$

the current wave is obtained in the form

$$\begin{aligned} i(x, t) = {} & i_0(t - \delta x)\Upsilon(t - \delta x) - \Gamma_l i_0[t - \delta(2l - x)]\Upsilon[t - \delta(2l - x)] \\ & + \Gamma_0\Gamma_l i_0[t - \delta(2l + x)]\Upsilon[t - \delta(2l + x)] \\ & - \Gamma_0\Gamma_l^2 i_0[t - \delta(4l - x)]\Upsilon[t - \delta(4l - x)] + \cdots \end{aligned} \tag{IV.15}$$

This expression represents the sum of a number of current waves, each of the same form as $i_0(t)$ [with the same remarks as in the case of $v(x, t)$ above]. The only fundamental difference between the current and voltage expressions is that, for the current, waves which have undergone a number of reflections at the output which is not equal to the number of reflections at the input appear with a minus sign.

2. *The input and output current and voltage*

Usually only the currents and voltages at the input or output of the line are of interest. For these quantities, it is useful to write down the simplest possible expressions.

(a) INPUT VOLTAGE AND CURRENT. Equation (IV.5) yields

$$V(0, p) = E(p)\frac{R_c}{Z_0 + R_c}(1 + \Gamma_l e^{-2p\tau})\sum_{n=0}^{\infty}\Gamma_0^n\Gamma_l^n e^{-2np\tau} \tag{IV.16}$$

In this form, $V(0, p)$ involves two series of delay functions,

$$\sum_{n=0}^{\infty} e^{-2np\tau}$$

and

$$\sum_{n=0}^{\infty} e^{-2(n+1)p\tau}$$

This corresponds to a double series of waves at the input. But the wave returning from the output interferes at the input with the wave which is returning to the output. It is desirable to consider only one series of waves. Accordingly, we start from Eq. (IV.3) to obtain

$$\begin{aligned} V(0, p) &= \frac{E(p)R_c}{Z_0 + R_c}\,\frac{1 + \Gamma_l e^{-2p\tau}}{1 - \Gamma_0\Gamma_l e^{-2p\tau}} \\ &= \frac{E(p)R_c}{Z_0 + R_c}\,\frac{1 + \dfrac{1}{\Gamma_0} - \dfrac{1}{\Gamma_0}(1 - \Gamma_0\Gamma_l e^{-2p\tau})}{1 - \Gamma_0\Gamma_l e^{-2p\tau}} \\ &= \frac{E(p)R_c}{Z_0 + R_c}\left[-\frac{1}{\Gamma_0} + \left(1 + \frac{1}{\Gamma_0}\right)\left(1 + \sum_{n=1}^{\infty}\Gamma_0^n\Gamma_l^n e^{-2np\tau}\right)\right] \end{aligned}$$

$$V(0, p) = \frac{E(p)R_c}{Z_0 + R_c}\left[1 + \left(1 + \frac{1}{\Gamma_0}\right)\sum_{n=1}^{\infty}\Gamma_0^n\Gamma_l^n e^{-2np\tau}\right] \tag{IV.17}$$

[To obtain this entirely in terms of Γ_0, $R_c/(Z_0 + R_c)$ can be replaced by $(1 - \Gamma_0)/2$.]

In the same way, it is possible to obtain

$$I(0, p) = \frac{E(p)}{Z_0 + R_c}\left[1 + \left(1 - \frac{1}{\Gamma_0}\right)\sum_{n=1}^{\infty}\Gamma_0^n\Gamma_l^n e^{-2pn\tau}\right] \tag{IV.18}$$

In case $Z_0 = R_0$ and $Z_l = R_l$, and letting $e(t)$ correspond to $E(p)$, these become†

† There should be no confusion between the generator voltage $e(t)$ and the number e, the base of the natural logarithms.

$$v(0, t) = \frac{R_c}{R_0 + R_c} \left[e(t)\, \Upsilon(t) + \left(1 + \frac{1}{\Gamma_0}\right) \sum_{n=1}^{\infty} \Gamma_0^n \Gamma_l^n e(t - 2n\tau)\, \Upsilon(t - 2n\tau) \right] \tag{IV.19}$$

$$i(0, t) = \frac{1}{R_0 + R_c} \left[e(t)\, \Upsilon(t) + \left(1 - \frac{1}{\Gamma_0}\right) \sum_{n=1}^{\infty} \Gamma_0^n \Gamma_l^n e(t - 2n\tau)\, \Upsilon(t - 2n\tau) \right] \tag{IV.20}$$

(b) Output Voltage and Current. For $x = l$, relations (IV.5) and (IV.6) of Section 2.2 become

$$V(l, p) = E(p) \frac{R_c}{Z_0 + R_c} (1 + \Gamma_l) \sum_{n=0}^{\infty} \Gamma_0^n \Gamma_l^n e^{-(2n+1)\tau p} \tag{IV.21}$$

$$I(l, p) = E(p) \frac{1}{Z_0 + R_c} (1 - \Gamma_l) \sum_{n=0}^{\infty} \Gamma_0^n \Gamma_l^n e^{-(2n+1)\tau p} \tag{IV.22}$$

In the case $Z_0 = R_0$ and $Z_l = R_l$, these become

$$v(l, t) = \frac{R_c}{R_0 + R_c} (1 + \Gamma_l) \sum_{n=0}^{\infty} \Gamma_0^n \Gamma_l^n e[t - (2n+1)\tau] \Upsilon[t - (2n+1)\tau] \tag{IV.23}$$

$$i(l, t) = \frac{1}{R_0 + R_c} (1 - \Gamma_l) \sum_{n=0}^{\infty} \Gamma_0^n \Gamma_l^n e[t - (2n+1)\tau] \Upsilon[t - (2n+1)\tau] \tag{IV.24}$$

3. *First example of multiple reflections*

Consider the circuit of Fig. 127. The generator delivers a step of voltage E at $t = 0$. The voltage and current as functions of time are to be found at the input, at the output, and at an arbitrary point x along the line.

In this example,

$$\Gamma_0 = \frac{R_0 - R_c}{R_0 + R_c} = \frac{25 - 50}{25 + 50} = -\frac{1}{3}$$

$$\Gamma_l = \frac{R_l - R_c}{R_l + R_c} = \frac{150 - 50}{150 + 50} = \frac{1}{2}$$

$$e(t) \frac{R_c}{R_0 + R_c} = E\, \Upsilon(t) \frac{50}{25 + 50} = \frac{2E}{3} \Upsilon(t)$$

$$e(t) \frac{1}{R_0 + R_c} = \frac{E}{75} \Upsilon(t)$$

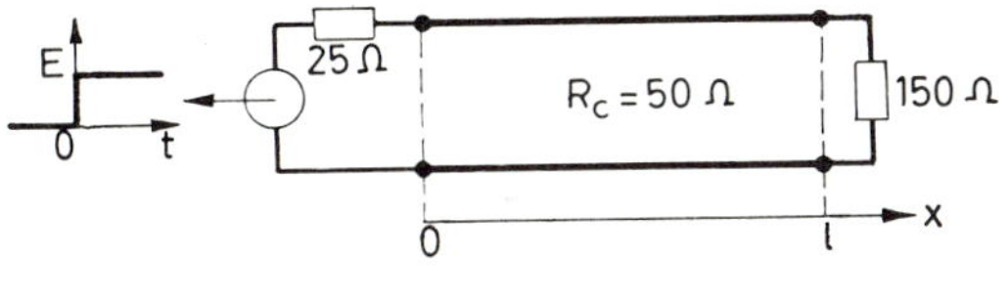

Fig. 127. Circuit with multiple reflections.

(a) THE INPUT. Using the values appropriate to this example, written above, in (IV.19) results in

$$v(0, t) = \frac{2E}{3}\left[\Upsilon(t) - 2\sum_{n=1}^{\infty}\left(-\frac{1}{3}\right)^n\left(\frac{1}{2}\right)^n \Upsilon(t - 2n\tau)\right]$$

$$= \frac{2E}{3}\left\{\Upsilon(t) + 2\left[\frac{1}{3 \times 2}\Upsilon(t - 2\tau) - \left(\frac{1}{3 \times 2}\right)^2 \Upsilon(t - 4\tau)\cdots\right]\right\}$$

To determine the final value of the voltage at the input, it is possible to sum the series

$$v(0, t=+\infty) = \frac{ER_c}{R_0 + R_c}\left[1 + \left(1 + \frac{1}{\Gamma_0}\right)(\Gamma_0\Gamma_l + \Gamma_0^2\Gamma_l^2 + \cdots + \Gamma_0^n\Gamma_l^n + \cdots)\right]$$

The result is

$$v(0, t = +\infty) = \frac{ER_l}{R_0 + R_l}$$

which is evidently correct. For this case,

$$v(0, t = +\infty) = E\frac{150}{150 + 25} = \frac{6}{7}E$$

From (IV.20), the current is

$$i(0, t) = \frac{E}{75}\left[\Upsilon(t) + 4\sum_{n=1}^{\infty}\left(-\frac{1}{3}\right)^n\left(\frac{1}{2}\right)^n \Upsilon(t - 2n\tau)\right]$$

$$= \frac{E}{75}\left\{\Upsilon(t) - 4\left[\frac{1}{3 \times 2}\Upsilon(t - 2\tau) - \left(\frac{1}{3 \times 2}\right)^2 \Upsilon(t - 4\tau) + \cdots\right]\right\}$$

The final value of the input current is

$$i(0, t = +\infty) = \frac{E}{R_0 + R_l} = \frac{E}{175}$$

(b) THE OUTPUT. Using the appropriate parameter values in (IV.23) yields

$$v(l, t) = \frac{2E}{3}\,\frac{3}{2}\sum_{n=0}^{\infty}\left(-\frac{1}{3}\right)^n\left(\frac{1}{2}\right)^n \Upsilon[t - (2n + 1)\tau]$$

$$= E\left[\Upsilon(t - \tau) - \frac{1}{3 \times 2}\Upsilon(t - 3\tau) + \left(\frac{1}{3 \times 2}\right)^2 \Upsilon(t - 5\tau) + \cdots\right]$$

The final value of the output voltage is

$$v(l, t = +\infty) = v(0, t = +\infty) = \frac{6E}{7}$$

From (IV.24), the current at the output is

$$i(l, t) = \frac{v(l, t)}{R_l}$$
$$= \frac{E}{150}\left[\Upsilon(t-\tau) - \frac{1}{3\times 2}\Upsilon(t-3\tau) + \left(\frac{1}{3\times 2}\right)^2 \Upsilon(t-5\tau) + \cdots\right]$$

(c) AN ARBITRARY POINT x. One could calculate $v(x, t)$ and $i(x, t)$ from (IV.14) and (IV.15), but it is much easier to deduce the voltage and current at a point x along the line from the voltage and current at the ends of the line, using the following remark. The voltage and current at a point x and at time t are equal to the voltage and current at the input at time $t - \delta x$, or at the output at time $t - \delta(l - x)$ (see Chapter II).

Thus $v(x, t)$, $i(x, t)$ can be found from $v(0, t)$, $i(0, t)$ or from $v(l, t)$, $i(l, t)$. The results thus obtained are plotted in Fig. 128 in normalized coordinates t/τ, $v(x, t)/E$, $i(x, t)/(E/75)$, for the particular choice $x = l/2$.

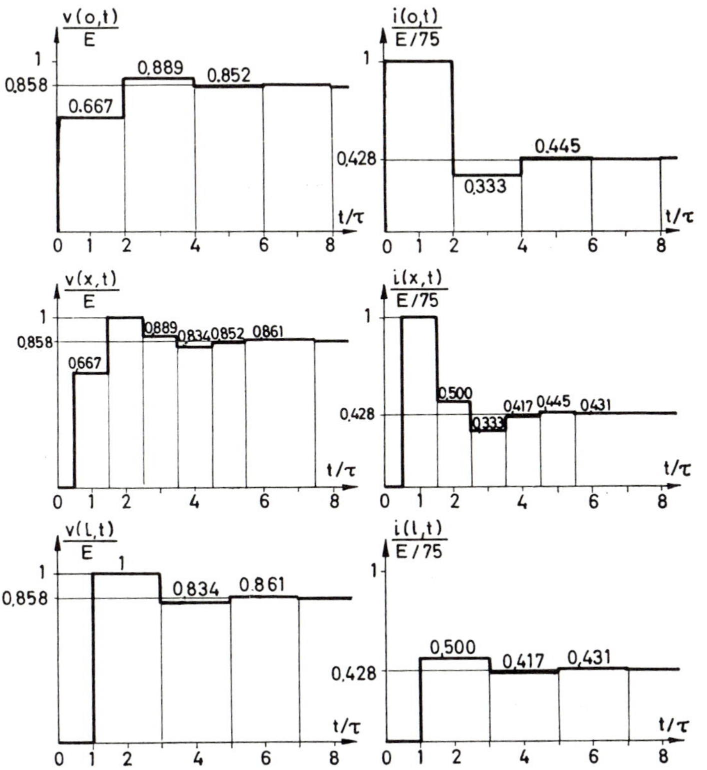

Fig. 128. Voltages and currents at the input, output, and in the middle of the line of Fig. 127.

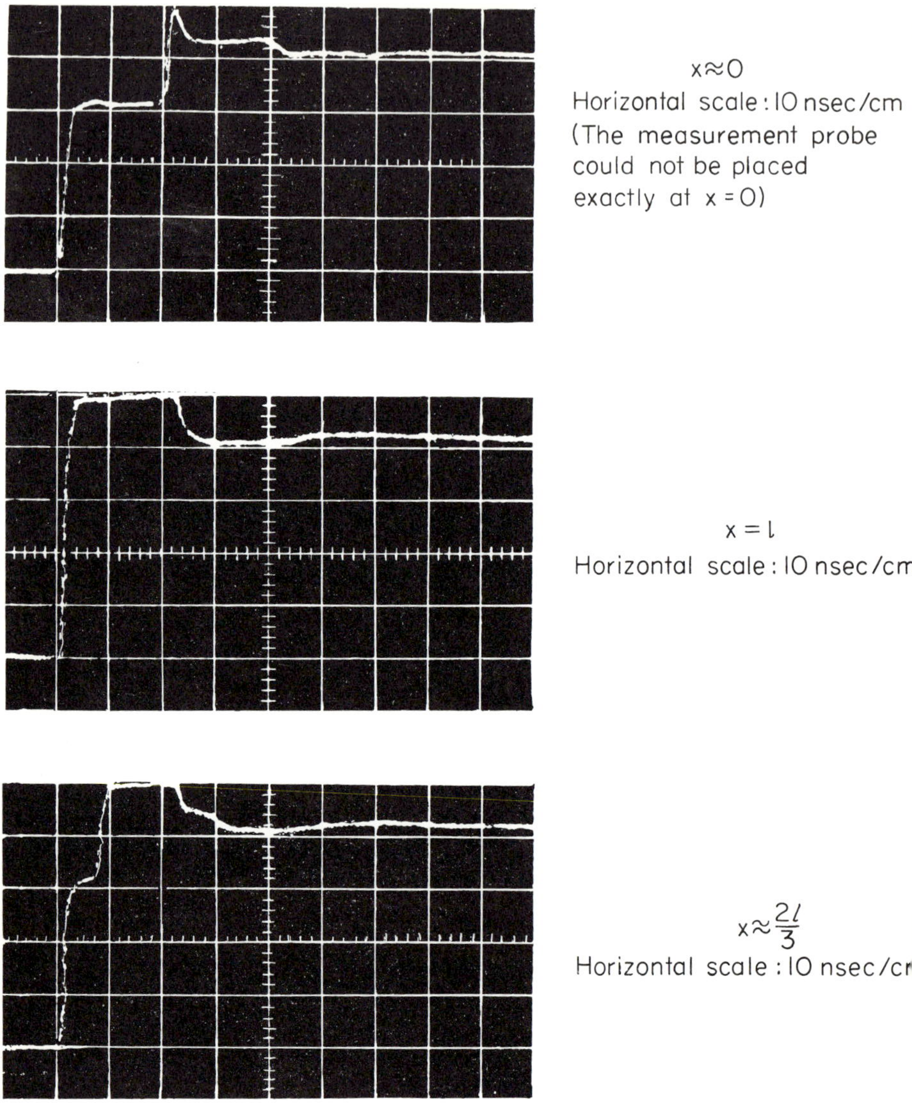

Fig. 129. Oscillograms of voltage waveforms along the line of Fig. 127.

Figure 129 shows oscilloscope photographs of the observed voltage at three points along a line of length 1.7 meters.

4. *Second example of multiple reflections*

Consider the circuit of Fig. 130. The output voltage $v(l, t)$ and the input current $i(0, t)$ are to be found for this open line.

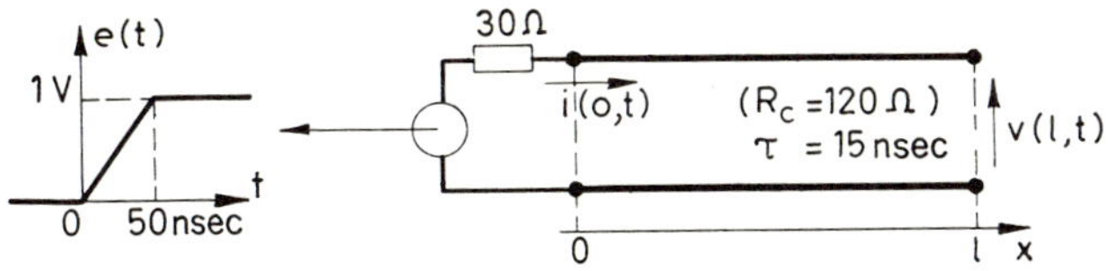

Fig. 130. Another line with multiple reflections present.

(a) Determination of $v(l, t)$. Using (IV.23)

$$v(l, t) = \frac{R_c}{R_0 + R_c}(1 + \Gamma_l)\sum_{n=0}^{\infty} \Gamma_0^n \Gamma_l^n e[t - (2n + 1)\tau]\Upsilon[t - (2n + 1)\tau]$$

with the parameters of this case

$$\Gamma_l = \frac{R_l - R_c}{R_l + R_c} = 1$$

$$\Gamma_0 = \frac{R_0 - R_c}{R_0 + R_c} = \frac{30 - 120}{30 + 120} = -\frac{3}{5}$$

$$\frac{R_c}{R_0 + R_c} = \frac{120}{30 + 120} = \frac{4}{5}; \qquad \tau = 15 \quad \text{nsec}$$

yields

$$v(l, t) = \frac{8}{5}\sum_{n=0}^{\infty}\left(-\frac{3}{5}\right)^n e[t - (2n + 1)\tau]\Upsilon[t - (2n + 1)\tau]$$

Thus the output voltage of the line is the sum of a series of voltage ramps, with amplitudes decreasing as $(-3/5)^n$, and beginning at times $t = \tau, 3\tau, \ldots, (2n + 1)\tau, \ldots$. The following table gives the characteristics of the first terms of the series:

n	$\frac{t}{\tau}$	*Amplitude* (V)
0	1	$8/5 = 1.60$
1	3	$8/5(-3/5) = -0.96$
2	5	$8/5(-3/5)^2 = 0.58$
3	7	$8/5(-3/5)^3 = -0.35$
4	9	$8/5(-3/5)^4 = 0.21$
5	11	$8/5(-3/5)^5 = -0.12$
6	13	$8/5(-3/5)^6 = 0.07$

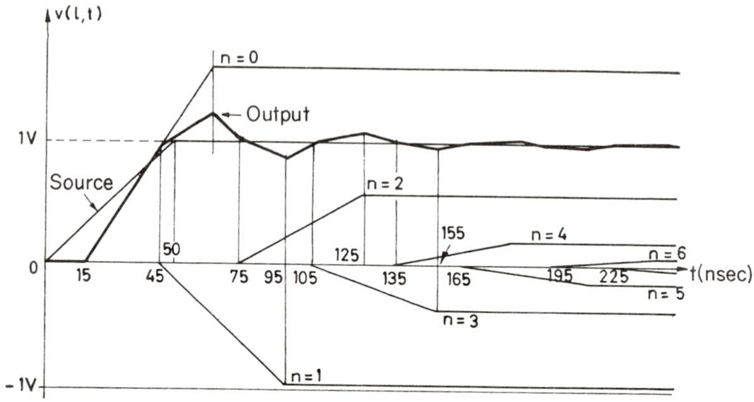

Fig. 131a. Output voltage for circuit of Fig. 130.

These ramps and their sum $v(l, t)$ are shown in Fig. 131a. Note the output voltage reaches 1 volt before the generator voltage reaches that level.

(b) DETERMINATION OF $i(0, t)$. Relation (IV.20)

$$i(0, t) = \frac{1}{R_0 + R_c} \left[e(t)\, \Upsilon(t) + \left(1 - \frac{1}{\Gamma_0}\right) \sum_{n=1}^{\infty} \Gamma_0^n \Gamma_l^n e(t - 2n\tau) \Upsilon(t - 2n\tau) \right]$$

with

$$1 - \frac{1}{\Gamma_0} = 1 + \frac{5}{3} = \frac{8}{3}$$

yields

$$i(0, t) = \frac{1}{R_0 + R_c} \left[e(t)\, \Upsilon(t) + \frac{8}{3} \sum_{n=1}^{\infty} \left(- \frac{3}{5} \right)^n e(t - 2n\tau) \Upsilon(t - 2n\tau) \right]$$

$$= \frac{1}{R_0 + R_c} \left[e(t)\, \Upsilon(t) - \frac{8}{5} \sum_{n=1}^{\infty} \left(- \frac{3}{5} \right)^{n-1} e(t - 2n\tau) \Upsilon(t - 2n\tau) \right]$$

Setting $n - 1 = m$, this becomes

$$i(0, t) = \frac{1}{R_0 + R_c} \left[e(t) \Upsilon(t) - \frac{8}{5} \sum_{m=0}^{\infty} \left(-- \frac{3}{5} \right)^m e[t - (2m+2)\tau] \Upsilon[t - (2m+2)] \right]$$

It is not necessary to calculate the amplitudes of the various ramps in this expression, since it contains $v(l, t)$, except for sign and an additional delay τ:

$$i(0, t) = \frac{1}{150} [e(t)\, \Upsilon(t) - v(l, t - \tau)]$$

The various ramps making up $i(0, t)$ and their sum are shown in Fig. 131b.

Remark. No particular difficulty arose in solving this example, even though the rise time of the generator ramp and the time of propagation along the line were not in a simple ratio. This would not have been the case if Bergeron's method had been used.

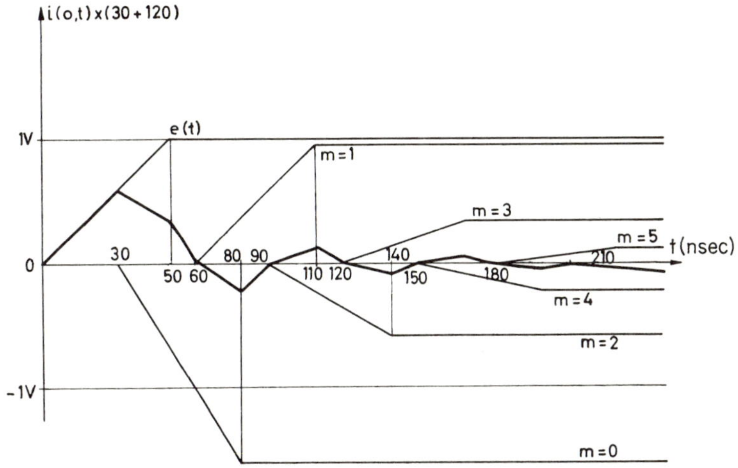

Fig. 131b. Output current for circuit of Fig. 130.

Generalization. If the generator voltage $e(t)$ is some function other than a ramp, the solution to the problem can again be obtained by graphical summation of delayed images of $e(t)$. It is only necessary to have a plot of $e(t)$, and it is not necessary to calculate $E(p)$.

2.8 *Conclusions concerning lossless lines*

The assumption of a lossless line has allowed a relatively simple solution for the transient behavior of the line to be found, and has allowed certain examples to be solved completely. In practice, however, lossless lines do not exist. Thus there will always be certain differences between the behavior encountered in practice, and that which has been calculated above. For example, infinitely steep wave fronts and square corners on the waveforms are never observed. The largest deviations from theory will occur at the passage of a wave front. It will next be shown how these wave fronts deteriorate in the presence of losses.

The study of transients on lossless lines can only give the general form of the true solution, but in many cases of practical interest, the lossless approximation suffices.

3. Lines with Loss in the Beginning of the Transient Regime

Experience shows that even in high-quality lines, losses deform the voltage and current wave shapes due to their effect on the amplitudes of the various waves making up the over-all line voltage and current. For this reason it is desirable to study the behavior of lossy lines. The region of greatest interest is the beginning of the transient regime, since the influence of successive waves becomes smaller and smaller, due to losses in the line and to reflection coefficients involving loss.

3.1 *Definition*

A time t is defined to be in the *beginning of the transient regime* provided $0 \leq t \ll t_1$, where t_1 is the smaller of L/R or C/G, L, R, C, and G being the per unit length parameters of the line.

For example, for RG8 coaxial cable,

$$\mathrm{L} = 2.5 \times 10^{-7}\ \mathrm{H/meter}; \qquad \mathrm{C} = 10^{-10}\ \mathrm{F/meter}; \qquad \mathrm{R} = 10^{-2}\ \Omega/\mathrm{meter};$$

$$\mathrm{G} = 0.6 \times 10^{-6}\ \Omega^{-1}/\mathrm{meter}$$

so that

$$\frac{\mathrm{L}}{\mathrm{R}} = 25\ \mu\mathrm{sec}; \qquad \frac{\mathrm{C}}{\mathrm{G}} = 160\ \mu\mathrm{sec}$$

Here the beginning of the transient regime might be taken as $0 \leq t \leq 5\ \mu$sec.

3.2 *Justification*

In the beginning of the transient regime, as defined above, the effects of series and shunt losses are very much less than the effects of inductance and capacitance. To see this it is only necessary to write the two conditions $t \ll \mathrm{L/R}$ and $t \ll \mathrm{C/G}$ in the Laplace transform domain:

$$\frac{1}{p^2} \ll \frac{\mathrm{L}}{\mathrm{R}p}; \qquad \frac{1}{p^2} \ll \frac{\mathrm{C}}{\mathrm{G}p}$$

to obtain

$$\mathrm{R} \ll \mathrm{L}p; \qquad \mathrm{G} \ll \mathrm{C}p$$

Remark. Such ordering relationships as these are valid only if p is real. Further, the inequality as we have treated it holds only for $p > 0$. Both these conditions are, however, compatible with the conditions for existence of the Laplace transforms of the functions with which we have been dealing.

3.3 *Approximate expressions for* Z_c

Developing Z_c in a series:

$$Z_c = \left(\frac{R + Lp}{G + Cp}\right)^{1/2} = \left(\frac{L}{C}\right)^{1/2}\left(\frac{1 + R/Lp}{1 + G/Cp}\right)^{1/2} = R_c\left[1 + \frac{1}{2p}\left(\frac{R}{L} - \frac{G}{C}\right) + \cdots\right]$$

yields the approximation

$$Z_c \approx R_c + \frac{1}{p}\left(\frac{R - GR_c^2}{2\delta}\right) \tag{IV.25}$$

Thus to the first order, for $0 < t \ll t_1$,

$$Z_c = R_c$$

since then $R \ll Lp$ and $G \ll Cp$, so that

$$\frac{1}{2p}\left(\frac{R}{L} - \frac{G}{C}\right) \ll 1$$

To the second order,

$$Z_c = R_c + \frac{1}{C_1 p}$$

where

$$C_1 = \frac{2\delta}{R - GR_c^2} > 0$$

Thus the characteristic impedance is equivalent to a resistance R_c in series with a capacitance C_1. This provides a means of matching the line, at least at the beginning of the transient regime.

Example. For RG8 coaxial cable, $R_c = 50\,\Omega$, $G = 0.6 \times 10^{-6}\,\Omega^{-1}$/meter, $R = 10^{-2}\,\Omega$/meter. Thus

$$(R - GR_c^2) = (10^{-2} - 0.6 \times 10^{-6} \times 50^2) \approx 10^{-2}\ (\Omega/\text{meter}); \quad C_1 = 1\ \mu\text{F}$$

Note that here $(R - GR_c^2)$ is indeed positive. This will always be the case.

3.4 *Approximate expressions for* γ

It was already mentioned in Chapter II that

$$\gamma = [(R + Lp)(G + Cp)]^{1/2} = p(LC)^{1/2}\left[\left(1 + \frac{R}{Lp}\right)\left(1 + \frac{G}{Cp}\right)\right]^{1/2}$$

$$= p\delta\left[1 + \frac{1}{2p}\left(\frac{R}{L} + \frac{G}{C}\right) + \cdots\right]$$

$$\gamma = p\delta + \frac{1}{2}\left(\frac{R}{R_c} + GR_c\right)$$

Thus to the first order, for $0 < t \ll t_1$,

$$\gamma = p\delta$$

since in that case $R \ll Lp$ and $G \ll Cp$, so that

$$\frac{1}{2p}\left(\frac{R}{L} + \frac{G}{C}\right) \ll 1$$

To the second order,

$$\gamma = p\delta + \alpha \tag{IV.26}$$

where

$$\alpha = \frac{1}{2}\left(\frac{R}{R_c} + GR_c\right) \tag{IV.27}$$

Here α is a real number, with the dimensions of an inverse length. It is the *attenuation constant*, and is measured in nepers per meter. Even if α is small with respect to $p\delta$, it should not be neglected, since its effect is quite different.

3.5 *Equations of the lossy line at the beginning of the transient regime*

On replacing γ by $(p\delta + \alpha)$ there results

$$\begin{aligned} V(x, p) &= E(p)\frac{Z_c}{Z_0 + Z_c}\,[e^{-(p\delta+\alpha)x} + \Gamma_l e^{-(p\delta+\alpha)(2l-x)}]\sum_{n=0}^{\infty}\Gamma_0^n\Gamma_l^n e^{-2n(p\delta+\alpha)l} \\ &= \frac{E(p)Z_c}{Z_0 + Z_c}\,[e^{-p\delta x}e^{-\alpha x} + \Gamma_l e^{-p\delta(2l-x)}e^{-\alpha(2l-x)}]\sum_{n=0}^{\infty}\Gamma_0^n\Gamma_l^n e^{-2np\delta l}e^{-2n\alpha l} \end{aligned} \tag{IV.28}$$

with a similar expression for the current $I(x, p)$. These contain delay factors $e^{-p\delta x}$ and $e^{-p\delta(2l-x)}$, so that, as for the lossless line, disturbances propagate with velocity $u = 1/\delta = 1/(LC)^{1/2}$. The factors $e^{-\alpha x}$ indicate that the amplitudes of the disturbances are attenuated by $e^{-\alpha}$ per unit length of line traversed. These results were used in Chapter II in the method of traveling waves.

Remark. Because of the attenuation, some lines can be considered to be of infinite length (i.e., matched at the output), since the waves which return to the origin after reflection may be of negligible amplitude.

3.6 *Example. Output voltage of an open lossy line excited by an ideal voltage step generator* (Fig. 132)

This problem was treated in Chapter II. For $Z_0 = 0$, $Z_l = \infty$, $\Gamma_0 = -1$, and $\Gamma_l = +1$, formula (IV.28) written for the output end of the line yields

$$V(l, p) = \frac{2E}{p} \sum_{n=0}^{\infty} (-1)^n e^{-(2n+1)\alpha l} e^{-(2n+1)\tau p}$$

$$v(l, t) = 2E \sum_{n=0}^{\infty} (-1)^n e^{-(2n+1)\alpha l} \Upsilon[t - (2n+1)\tau]$$

$$v(l, t) = + 2Ee^{-\alpha l}\Upsilon(t - \tau) - 2Ee^{-3\alpha l}\Upsilon(t - 3\tau) + 2Ee^{-5\alpha l}\Upsilon(t - 5\tau) - 2Ee^{-7\alpha l}\Upsilon(t - 7\tau) + \cdots$$

Fig. 132. Open lossy line driven by ideal source.

Thus the output voltage is the sum of terms which are alternately positive and negative, and which are of decreasing amplitude. The output voltage of the line is shown in Fig. 133a, along with the case of a lossless line for comparison. Figure 133b shows an actual oscilloscope photograph of the voltage at the output of a 50-meter length of RG8 cable. The progressive

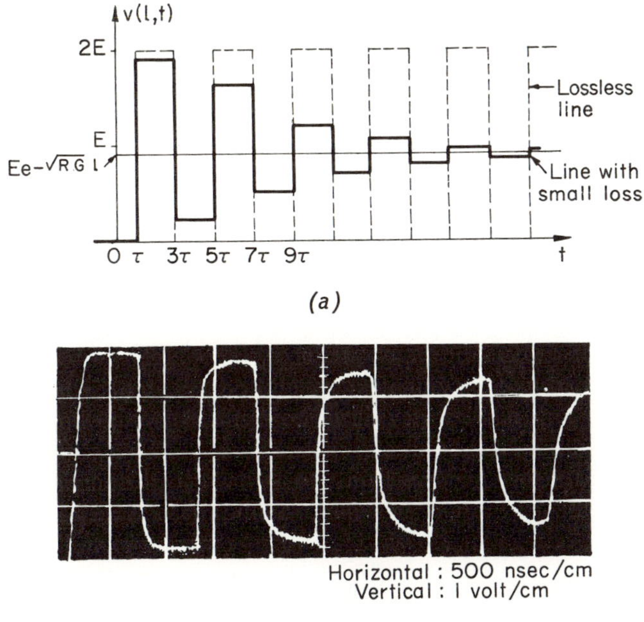

Fig. 133. Output voltage for circuit of Fig. 132. (a) Theoretical waveform. (b) Oscillogram.

decrease in amplitude is apparent. The progressive deformation of the waveform is due to skin effect (see Section 5 below).

Remark. Let us find the steady-state output current and voltage for this case. The infinite series of terms obtained for $t = \infty$,

$$v(l, t = \infty) = 2E(e^{-\alpha l} - e^{-3\alpha l} + e^{-5\alpha l} - e^{-7\alpha l} + \cdots)$$

could be summed to obtain $(\cosh \alpha l)/E$, but this is not valid, since for large t, we are no longer in the beginning of the transient regime.

To determine the correct steady-state conditions, it is necessary to return to the basic relations of Chapter I. For the steady state, recall that

$$\frac{\partial i}{\partial t} = \frac{\partial v}{\partial t} = 0$$

so that L and C have no effect. There remains

$$\frac{\partial^2 v}{\partial x^2} = \mathrm{RG}v$$

from which

$$v(x, t = \infty) = \mathrm{A}\exp[-(\mathrm{RG})^{1/2}x] + \mathrm{B}\exp[+(\mathrm{RG})^{1/2}x]$$

Setting $x = \infty$ in this, $\mathrm{B} = 0$ results, since the voltage can not be infinite. Setting $x = 0$ results in $\mathrm{A} = \mathrm{E}$. Thus the output voltage is

$$v(l, t = \infty) = \mathrm{E}\exp[-(\mathrm{RG})^{1/2}l] \tag{IV.29}$$

Also,

$$\frac{\partial i}{\partial x} = -\mathrm{G}v = -\mathrm{GE}\exp[-(\mathrm{RG})^{1/2}x] \quad \text{for} \quad t \to +\infty$$

which when integrated results in

$$i(x, t = \infty) = \frac{\mathrm{G}}{(\mathrm{RG})^{1/2}}\mathrm{E}\exp[-(\mathrm{RG})^{1/2}x] + \mathrm{K}$$

For $x = l$, $i = 0$, which allows K to be found. The input current which results is

$$i(0, t = \infty) = (\mathrm{G/R})^{1/2}\mathrm{E}(1 - \exp[-(\mathrm{RG})^{1/2}l]) \tag{IV.30}$$

4. Lossy Lines Satisfying the Heaviside Condition

This type of line is now only of historical interest. The *Heaviside condition* is $\mathrm{R/L} = \mathrm{G/C}$ (whereas for a real line $\mathrm{R/L} > \mathrm{G/C}$). Under this condition,

$$\mathrm{Z}_c = \left(\frac{\mathrm{R} + \mathrm{L}p}{\mathrm{G} + \mathrm{C}p}\right)^{1/2} = \left(\frac{\mathrm{L}}{\mathrm{C}}\right)^{1/2}\left(\frac{\mathrm{R/L} + p}{\mathrm{G/C} + p}\right)^{1/2} = \left(\frac{\mathrm{L}}{\mathrm{C}}\right)^{1/2} = \mathrm{R}_c \tag{IV.31}$$

and the characteristic impedance is a pure resistance R_c, just as if the line were lossless. Further,

$$\begin{aligned}\gamma &= [(R + Lp)(G + Cp)]^{1/2} = p(LC)^{1/2}(1 + R/Lp) \\ &= p\delta + R(C/L)^{1/2} = p\delta + R/R_c = \delta(p + R/L)\end{aligned} \tag{IV.32}$$

These formulas are rigorous, and involve no approximations. Thus disturbances propagate along such a line with velocity $u = 1/(LC)^{1/2}$, and undergo attenuation of $\exp(-R/R_c)$ per unit length. Since

$$\left(\frac{R}{R_c}\right) > \frac{1}{2}\left(\frac{R}{R_c} + GR_c\right) = \alpha \tag{IV.33}$$

(because most often $R/R_c > GR_c$, so that $L/R < G/C$), the attenuation of a line satisfying the Heaviside condition is greater than that of a lossy line in the beginning of the transient regime.

To realize a line satisfying the Heaviside condition, the shunt loss can be artificially increased (thus increasing the total loss), or L can be increased artificially.

The Heaviside condition is largely of theoretical interest. To our knowledge, this condition is no longer satisfied in existing transmission lines. In addition, since the line parameters are functions of frequency, as discussed in Chapter I, it is not possible to satisfy this condition over a frequency band sufficiently wide as to allow proper transmission of short pulses, in the range of 1 to 5 nsec, for example. On the other hand, the situation is different if only operation at a fixed frequency is of interest. The inductance per unit length, L, can then be easily increased so as to satisfy the Heaviside condition. In the past this led to two techniques used in telephony: lumped loading ("Pupinization"), in which lumped inductances are inserted at discrete points along the line, and uniform loading ("Krarupization"), in which magnetic material formed into a long tape is wound around the line throughout its length. In both cases, the propagation velocity $u = 1/(LC)^{1/2}$ decreases. Carrier telephony and the use of line repeaters have lately rendered even these artifices obsolete.

5. The Influence of Skin Effect†

5.1 *Introduction*

In all cases considered thus far (lossless lines, lines at the beginning of the transient regime, lines satisfying the Heaviside condition), the voltage

† See the following two articles:

R. L. Wigington and N. S. Nahman, Transient analysis of coaxial cables considering skin effect, *Proc. IRE*, February (1957).

at the output of a matched line had the same waveform as that at the input. Theoretically, all these lines are distortionless. This is a consequence of the assumption that all the line parameters were constant. But in practice there is noticeable deterioration of the voltage wave fronts, on a nanosecond scale, for propagation along cables only several meters long. This deterioration is principally due to skin effect, that is, to the fact that in the beginning of the transient regime, currents only flow in a thin layer on the surface of the conductors.

In this section we will study the response of a cable to a voltage step, taking account of skin effect, in the case that shunt losses can be neglected. In this case it is possible to define an operational resistance per unit length $R(p) = Kp^{1/2}$, where K is independent of p, but a function of the geometric and electromagnetic parameters of the cable (see Appendix to the present chapter). This relation only holds if the radius of curvature of a cross section of one of the conductors of the line is so large that it can be considered a plane conductor. This is substantially the case in practice (see Section 5.4).

5.2 *Response to a voltage step*

Assume that at $t = 0$ there is applied to a line of length l, terminated in its characteristic impedance, a constant voltage E (Fig. 134). The basic relation (IV.3) in this case ($\Gamma_0 = -1$, $\Gamma_l = 0$) yields for $x = l$

$$V(l, p) = \frac{E}{p} \exp[-\gamma(p)l] \tag{IV.34}$$

But with $R = Kp^{1/2}$ and $G = 0$,

$$\begin{aligned}\gamma(p) &= [(pL + R)(pC + G)]^{1/2} \\ &= [(pL + Kp^{1/2})pC]^{1/2} = p(LC)^{1/2}\left(1 + \frac{K}{Lp^{1/2}}\right)^{1/2}\end{aligned}$$

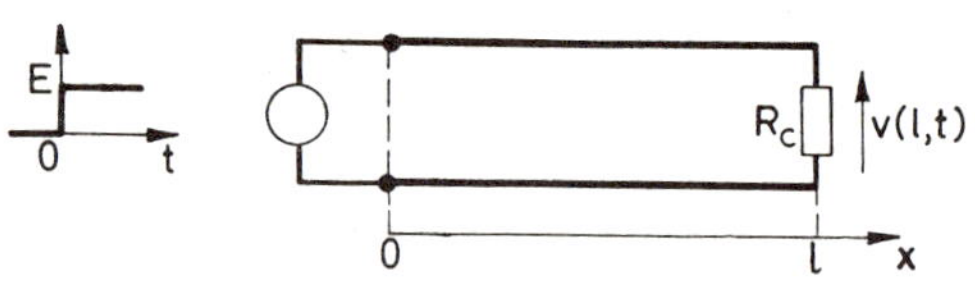

Fig. 134. Line with matched load and ideal source.

Let us suppose that $K/(Lp^{1/2}) \ll 1$, which is to say the line is in the beginning of the transient regime (see Section 5.4 for the conditions under

Q. Kerns, F. Kirsten, and C. Winningstadt, Pulse response of coaxial cables, Counting note File No. CC2-1, Radiation Laboratory, University of California, Berkeley.

which this is the case). Since we are primarily interested in affairs at times near the time at which the wave front reaches the output of the line, it suffices to retain the first two terms in the series:

$$\left(1+\frac{\mathrm{K}}{\mathrm{L}p^{1/2}}\right)^{1/2}\approx 1+\frac{\mathrm{K}}{2\mathrm{L}p^{1/2}}$$

Then

$$\gamma(p)=p(\mathrm{LC})^{1/2}\left(1+\frac{\mathrm{K}}{2\mathrm{L}p^{1/2}}\right)=p(\mathrm{LC})^{1/2}+\frac{\mathrm{K}}{2}\left(\frac{\mathrm{C}}{\mathrm{L}}\right)^{1/2}p^{1/2}=p\delta+\frac{\mathrm{K}}{2\mathrm{R}_c}p^{1/2} \tag{IV.35}$$

and relation (IV.34) becomes finally

$$\mathrm{V}(l,p)=\frac{\mathrm{E}}{p}\exp\left[-\left(p\delta+\frac{\mathrm{K}}{2\mathrm{R}_c}p^{1/2}\right)l\right]=\mathrm{E}\,e^{-p\tau}\,\frac{\exp[-(\mathrm{K}l/2\mathrm{R}_c)p^{1/2}]}{p}$$

The term $\exp(-p\tau)$ indicates the delay due to the propagation time of the cable. From transform tables, the inverse of the term $(1/p)(\exp-\theta p^{1/2})$ is $1-\mathrm{erf}(\theta/2t^{1/2})$, where $\mathrm{erf}(\theta/2t^{1/2})$ is the error function of $\mathrm{X}=\theta/2t^{1/2}$, that is,

$$\frac{2}{\pi^{1/2}}\int_0^x e^{-u^2}\,du$$

Thus the voltage $v(l,t)$ at the output of the line is

$$v(l,t)=\mathrm{E}\left[1-\mathrm{erf}\left(\frac{\mathrm{K}l}{4\mathrm{R}_c(t-\tau)^{1/2}}\right)\right]\Upsilon(t-\tau)$$

or, shifting the time origin to the instant τ at which the wave front arrives at the output,

$$v(l,t)=\mathrm{E}\left[1-\mathrm{erf}\left(\frac{\mathrm{K}l}{4\mathrm{R}_c t^{1/2}}\right)\right]\Upsilon(t)$$

$$=\mathrm{E}\left[1-\mathrm{erf}\left(\frac{1}{2}\left(\frac{\mathrm{K}^2l^2}{4\,\mathrm{R}_c^2t}\right)^{1/2}\right)\right]\Upsilon(t)$$

The term $\mathrm{K}^2l^2/4\mathrm{R}_c^2$ has the dimension of time, and since $\mathrm{erf}(\frac{1}{2})\approx\frac{1}{2}$, we are led to define

$$\mathrm{K}^2l^2/4\mathrm{R}_c^2=\mathrm{T}_{0.5} \tag{IV.36}$$

where $\mathrm{T}_{0.5}$ is then the time required for the output voltage to reach E/2. The above expression for $v(l,t)$ then becomes

$$v(l,t)=\mathrm{E}\left[1-\mathrm{erf}\left(\frac{1}{2}\left(\frac{\mathrm{T}_{0.5}}{t}\right)^{1/2}\right)\right]\Upsilon(t) \tag{IV.37}$$

The "time constant" $T_{0.5}$ is a fixed parameter for a given length of a given type of cable. It increases as the square of the length of the cable. Through K and R_c it is related to the dimensions of the line and to the conductor material.

Figure 135 shows in normalized coordinates $t/T_{0.5}$ and $v(l, t)/E$, the response of a line to a voltage step, taking account of skin effect, according to relation (IV.37). The output voltage reaches 0.50 E after a time $T_{0.5}$, reaches 0.90 E after 30 $T_{0.5}$, and reaches 0.95 E after 110 $T_{0.5}$.

5.3 *Expression of* $T_{0.5}$ *as a function of losses*

Recall expression (IV.35) for the propagation function:

$$\gamma(p) = p\delta + \frac{K}{2R_c} p^{1/2}$$

The expression we are seeking will result by considering the line in the sinusoidal steady state, replacing p by $j\omega$:

$$\gamma(j\omega) = j\omega\delta + \frac{K}{2R_c}(j\omega)^{1/2} = j\omega\delta + \frac{K}{2R_c}\omega^{1/2}e^{j(\pi/4)}$$

$$= \frac{K}{2R_c}\left(\frac{\omega}{2}\right)^{1/2} + j\left(\omega\delta + \frac{K}{2R_c}\left(\frac{\omega}{2}\right)^{1/2}\right)$$

Clearly $(K/2R_c)(\omega/2)^{1/2} = (K/2R_c)(\pi f)^{1/2}$ represents the attenuation α per unit length (in nepers per meter) at the frequency f. Thus

$$\alpha = \frac{K}{2R_c}(\pi f)^{1/2}$$

$$\frac{K^2 l^2}{4R_c^2} = \frac{\alpha^2 l^2}{\pi f} = T_{0.5}$$

$$T_{0.5} = \frac{\alpha^2 l^2}{\pi f} \qquad \text{(IV.38)}$$

where αl is the total attenuation from input to output. Thus to calculate the time constant $T_{0.5}$, it is only necessary to know the attenuation α of the cable at the frequency f of interest.

The measurement of $T_{0.5}$ based on losses (measurement of αl), can be made at any frequency at which skin effect is present. In practice, a frequency near but lower than the highest frequency in the spectra of the pulses of interest is used. For example, if pulses having rise times of the order of 1 nsec are of interest, a frequency of 1000 MHz might be used. This is the procedure used in Section 5.5 below.

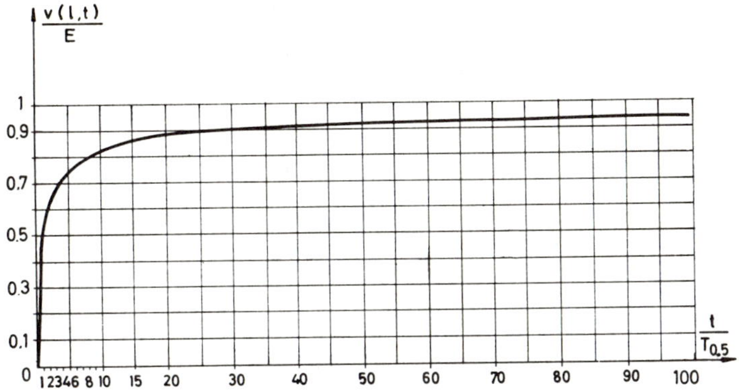

Fig. 135. Universal response curve for the line of Fig. 134, taking account of skin effect.

5.4 *Validity of* $K/(Lp^{1/2}) \ll 1$ *(beginning of the transient regime)*

Taking p to be real and positive (which is compatible with the defining conditions for the Laplace transform), we have

$$\frac{K}{L}\,\frac{1}{p} \ll \frac{p^{1/2}}{p}$$

But $p^{1/2}/p$ is the transform of $1/(\pi t)^{1/2}$, so that taking the inverse transform of the above inequality yields $K/L \ll 1/(\pi t)^{1/2}$, or $t \ll L^2/\pi K^2$. Thus we have

$$\frac{K}{Lp^{1/2}} \ll 1$$

for

$$0 < t \ll \frac{L^2}{\pi K^2}$$

The duration of the beginning of the transient regime is thus independent of the length of the cable.

The time constant $T_{0.5}$ and the duration t of the length of the transient regime can be related as follows. We have

$$T_{0.5} = \frac{K^2 l^2}{4R_c^2}$$

from which

$$\frac{1}{K^2} = \frac{l^2}{4R_c^2 T_{0.5}}$$

Thus

$$t \ll \frac{L^2 l^2}{4\pi R_c^2 T_{0.5}} = \frac{LCl^2}{4\pi T_{0.5}}$$

Writing $(LC)^{1/2} l = \tau$, this becomes

$$t \ll \frac{\tau^2}{4\pi T_{0.5}}$$

Since t is independent of the length of the cable, while $T_{0.5}$ is proportional to the length squared, the ratio $t/T_{0.5}$ is the smaller the larger is the length l.

5.5 *Numerical example for RG8 coaxial cable*

Let the length be 25 meters and the characteristic resistance be $R_c = 50\ \Omega$. The other characteristics are: inside radius, $r_1 = 1$ mm $= 10^{-3}$ meter; outside radius, $r_2 = 3.55$ mm $= 3.55 \times 10^{-3}$ meter. For copper, which we assume, $\mu = \mu_0 = 4\pi \times 10^{-7}$ H/meter, $\sigma = 0.6 \times 10^8\ (\Omega\text{meter})^{-1}$.

The time constant $T_{0.5}$ can be deduced from measurement of the loss of the line. At 1000 MHz an attenuation of 7.5 dB is measured, so that $\alpha l = 0.86$ Np. Then $T_{0.5} = (0.86)^2/(\pi \times 10^9) = 0.24 \times 10^{-9}$ sec, and $110\ T_{0.5} = 26$ nsec. Figure 136 shows the response observed using a sam-

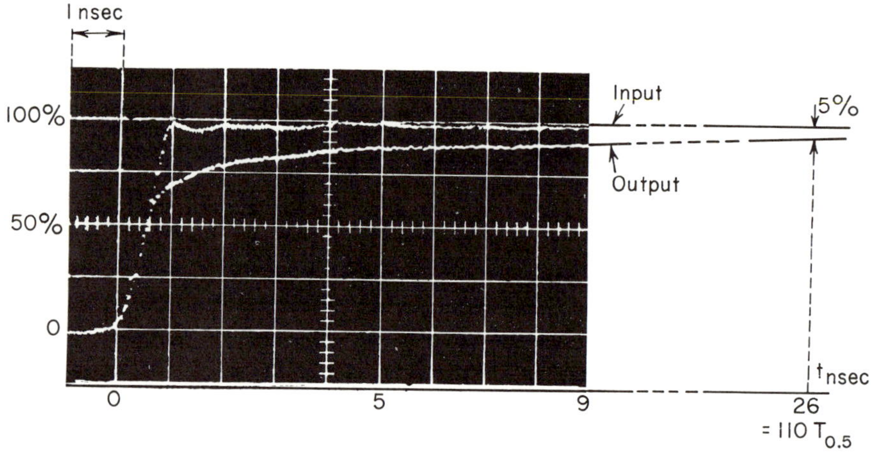

Fig. 136. Typical wavefront distortion caused by skin effect.

pling oscilloscope with 0 to 100% rise time of the order of 1 nsec. Here $T_{0.5}$ is not observable, being a good deal less than the rise time of the oscilloscope. Also $110T_{0.5}$ (26 nsec) is off the scale of the screen. But the figure does serve to show the degradation of wave front due to skin effect, and its characteristic form.

Let us check the inequality $K/Lp^{1/2} \ll 1$. Measurement of the cable loss led to a value $T_{0.5} = 0.24$ nsec. In Section 2.4 we saw that the delay was 5 nsec per meter of cable, which gives a propagation time $\tau = 5 \times 25 = 125$ nsec. Then we have

$$t \ll \frac{\tau^2}{4\pi T_{0.5}} = \frac{(125 \times 10^{-9})^2}{4\pi \times 0.25 \times 10^{-9}} \approx 5 \quad \mu\text{sec}$$

For the output voltage to reach 95% of its final value E requires a time $110T_{0.5}$, or 26 nsec. Added to the transmission time τ, the result is still much less than 5 μsec. Thus the approximation $K/Lp^{1/2} \ll 1$ holds true.

6. An Equivalent Lumped Circuit for a Line†

6.1 *Introduction*

Consider a section of line with parameters R_c, τ, connecting a voltage generator with internal resistance R_0 to a resistance R_l. The purpose of the circuit is to transmit as quickly as possible, a given current I to the load R_l. To this end, the generator supplies a voltage step of amplitude $E = I(R_0 + R_c)$. A transmission line is necessary because the generator and the load are not located at the same place. The resistances R_0 and R_l are usually so small that $R_0 < R_c$ and $R_l < R_c$. Under these conditions, the current is established in the line by multiple reflections at the two ends, each reflection resulting in an increase of current. The final current is established as the result of an infinite number of step increases, of decreasing amplitude. The method of Bergeron provides a simple and rapid graphical solution to the problem, and the use of operational calculus allows the solution to be computed with any desired precision.

In many practical cases, however, the precision and detail of solution provided by the above two methods are not necessary. It is then the usual practice to replace the line by an equivalent inductance $L_{eq} = \tau R_c$. This substitution can be justified intuitively by the fact that only currents are of interest, and the line in effect restrains the establishment of the current, there being a progressive growth of current without oscillation around the final value. The simplified circuit of Fig. 137 is thus substituted for the actual line, for purposes of analysis.

If a graph of the evolution of the current at the middle of the line is compared to the current in the equivalent inductance, it is found that to a first approximation these currents are equal for all multiples of the prop-

† After J. Hug.

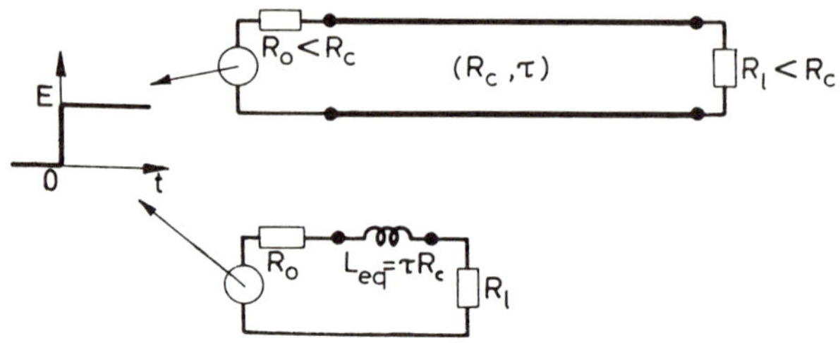

Fig. 137. Simple equivalent circuit for a line with small source and load resistances.

agation time τ of the line (Fig. 139). We will now give a theoretical justification for this observation, and for the approximation which it implies, assuming that the line is lossless.

6.2 *Currents at the middle of the line and in the equivalent inductance*

Since only conditions at the middle of the line are of interest, $x = l/2$ and $\delta x = \tau/2$. The input and output reflection coefficients are

$$\Gamma_0 = \frac{R_0 - R_c}{R_0 + R_c} < 0; \qquad \Gamma_l = \frac{R_l - R_c}{R_l + R_c} < 0$$

Let us define

$$\Gamma = \Gamma_0 \Gamma_l = \frac{(R_0 - R_c)\,(R_l - R_c)}{(R_0 + R_c)\,(R_l + R_c)}$$

(Γ is positive).

The relation giving the current at the middle of the line is [see (IV.15)]

$$i\left(\frac{l}{2}, t\right) = \frac{E}{R_c + R_0} \sum_{n=0}^{\infty} \Gamma^n \Upsilon[t - (2n + \tfrac{1}{2})\tau] - \frac{E\Gamma_l}{R_c + R_0} \sum_{n=0}^{\infty} \Gamma^n \Upsilon[t - (2n + \tfrac{3}{2})\tau]$$

or

$$i\left(\frac{l}{2}, t\right) = \frac{E}{(R_c + R_0)(R_c + R_l)} \times \left[(R_c + R_l) \sum_{n=0}^{\infty} \Gamma^n \Upsilon[t - (2n + \tfrac{1}{2})\tau] + (R_c - R_l) \sum_{n=0}^{\infty} \Gamma^n \Upsilon[t - (2n + \tfrac{3}{2})\tau]\right]$$

Using the normalized time t/τ and the normalized current at the middle of the line

$$\mathrm{I_L}\left(\frac{t}{\tau}\right) = \frac{i(l/2, t/\tau)}{\mathrm{E}/(\mathrm{R}_0 + \mathrm{R}_l)}$$

we have

$$\mathrm{I_L}\left(\frac{t}{\tau}\right) = \frac{\mathrm{R}_0 + \mathrm{R}_l}{(\mathrm{R}_c + \mathrm{R}_0)(\mathrm{R}_c + \mathrm{R}_l)} \times \left[(\mathrm{R}_c + \mathrm{R}_l) \sum_{n=0}^{\infty} \Gamma^n \Upsilon\left[\frac{t}{\tau} - (2n + \tfrac{1}{2})\right] + (\mathrm{R}_c - \mathrm{R}_l) \sum_{n=0}^{\infty} \Gamma^n \Upsilon\left[\frac{t}{\tau} - (2n + \tfrac{3}{2})\right]\right] \tag{IV.39}$$

The current $i_{\mathrm{S}}(t)$ in the self-inductance is

$$i_{\mathrm{S}}(t) = \frac{\mathrm{E}}{\mathrm{R}_0 + \mathrm{R}_l}\left(1 - \exp\left(-\frac{t}{\tau}\frac{\mathrm{R}_0 + \mathrm{R}_l}{\mathrm{R}_c}\right)\right)$$

Introducing the normalized current

$$\mathrm{I_S}\left(\frac{t}{\tau}\right) = \frac{i_{\mathrm{S}}(t/\tau)}{\mathrm{E}/(\mathrm{R}_0 + \mathrm{R}_l)}$$

this becomes

$$\mathrm{I_S}\left(\frac{t}{\tau}\right) = 1 - \exp\left(-\frac{t}{\tau}\frac{\mathrm{R}_0 + \mathrm{R}_l}{\mathrm{R}_c}\right) \tag{IV.40}$$

6.3 *Comparison of* $\mathrm{I_L}$ *and* $\mathrm{I_S}$ *for* $t/\tau = 2q$, *where* q *is an integer*

For $t/\tau = 2q$,

$$\mathrm{I_S}(2q) = 1 - \exp\left(-2q\,\frac{\mathrm{R}_0 + \mathrm{R}_l}{\mathrm{R}_c}\right)$$

To calculate $\mathrm{I_L}(2q)$ it is first necessary to determine the maximum value of n necessary in relation (IV.39). The term $\Upsilon((t/\tau) - (2n + \frac{1}{2}))$ is effective for $2q \geq 2n + \frac{1}{2}$, from which $n \leq q - \frac{1}{4}$, or $n_{\max} = q - 1$. The term $\Upsilon((t/\tau) - (2n + \frac{3}{2}))$ is effective for $2q \geq 2n + \frac{3}{2}$, from which $n \leq q - \frac{3}{4}$, or $n_{\max} = q - 1$. Hence

$$\begin{aligned}\mathrm{I_L}(2q) &= \frac{\mathrm{R}_0 + \mathrm{R}_l}{(\mathrm{R}_c + \mathrm{R}_0)(\mathrm{R}_c + \mathrm{R}_l)} \times [(\mathrm{R}_c + \mathrm{R}_l)(1 + \Gamma + \cdots + \Gamma^{q-1}) + (\mathrm{R}_c - \mathrm{R}_l)(1 + \Gamma + \cdots + \Gamma^{q-1})\\ &= \frac{2\mathrm{R}_c(\mathrm{R}_0 + \mathrm{R}_l)}{(\mathrm{R}_c + \mathrm{R}_0)(\mathrm{R}_c + \mathrm{R}_l)}(1 + \Gamma + \cdots + \Gamma^{q-1})\\ &= \frac{2\mathrm{R}_c(\mathrm{R}_0 + \mathrm{R}_l)}{(\mathrm{R}_c + \mathrm{R}_0)(\mathrm{R}_c + \mathrm{R}_l)}\,\frac{1 - \Gamma^q}{1 - \Gamma} = 1 - \Gamma^q\end{aligned}$$

Thus

$$\begin{aligned} I_L(2q) - I_S(2q) &= \exp\left(-2q\,\frac{R_0 + R_l}{R_c}\right) - \Gamma^q \\ &= \exp\left(-2q\,\frac{R_0 + R_l}{R_c}\right) - \exp(\mathrm{Log}\,\Gamma^q) \\ &= \exp\left(-2q\,\frac{R_0 + R_l}{R_c}\right) - \exp\left(-q\,\mathrm{Log}\frac{1}{\Gamma}\right) \end{aligned}$$

To compare $I_L(2q)$ to $I_S(2q)$, it is only necessary to compare $-2q(R_0 + R_l)/R_c$ to $-q\,\mathrm{Log}(1/\Gamma)$, or simply to compare $2(R_0 + R_l)/R_c$ with $\mathrm{Log}(1/\Gamma)$.

To simplify the notation, let $R_0/R_c = r_0$ and $R_l/R_c = r_l$. Then to compare $2(r_0 + r_l)$ to $\mathrm{Log}(1/\Gamma)$, we consider

$$\mathrm{Log}\,\frac{1}{\Gamma} = \mathrm{Log}\left[\frac{(R_c + R_0)}{(R_c - R_0)}\,\frac{(R_c + R_l)}{(R_c - R_l)}\right] = \mathrm{Log}\left(\frac{1 + r_0}{1 - r_0}\right) + \mathrm{Log}\left(\frac{1 + r_l}{1 - r_l}\right)$$

If r_0 and r_l are $\ll 1$, which is to say $R_0/R_c \ll 1$ and $R_l/R_c \ll 1$, then the two Log functions can be replaced by the first few terms of their series expansions. Specifically, we use

$$\mathrm{Log}\left(\frac{1 + r_0}{1 - r_0}\right) \approx 2\left(r_0 + \frac{r_0^3}{3}\right); \qquad \mathrm{Log}\left(\frac{1 + r_l}{1 - r_l}\right) \approx 2\left(r_l + \frac{r_l^3}{3}\right)$$

so that

$$\mathrm{Log}\,\frac{1}{\Gamma} \approx 2(r_0 + r_l) + \frac{2}{3}\,(r_0^3 + r_l^3) > 2(r_0 + r_l)$$

In summary, if $R_0/R_c \ll 1$ and $R_l/R_c \ll 1$, then for any q,

$$I_L(2q) \approx I_S(2q), \quad \text{but} \quad I_L(2q) > I_S(2q) \tag{IV.41}$$

6.4 *Comparison of* I_L *and* I_S *for* $t/\tau = 2q + 1$, *q an integer*

For $t/\tau = 2q + 1$,

$$I_S(2q + 1) = 1 - \exp[-(2q + 1)(r_0 + r_l)]$$

It is again necessary to determine the maximum value of n which need be considered in each of the step functions of (IV.39). For $\Upsilon(t/\tau - (2n + \frac{1}{2}))$ to be effective it is necessary that $2q + 1 \geq 2n + \frac{1}{2}$, from which $n \leq q + \frac{1}{4}$, or $n_{\max} = q$. For $\Upsilon(t/\tau - (2n + \frac{3}{2}))$ to be effective, it is necessary that $2q + 1 \geq 2n + \frac{3}{2}$, from which $n \leq q - \frac{1}{4}$, or $n_{\max} = q - 1$. Hence

$$
\begin{aligned}
\mathrm{I_L}(2q+1) &= \mathrm{I_L}(2q) + \frac{\mathrm{R}_0 + \mathrm{R}_l}{(\mathrm{R}_c + \mathrm{R}_0)(\mathrm{R}_c + \mathrm{R}_l)} (\mathrm{R}_c + \mathrm{R}_l)\Gamma^q \\
&= 1 - \Gamma^q + \frac{\mathrm{R}_0 + \mathrm{R}_l}{\mathrm{R}_c + \mathrm{R}_0} \Gamma^q \\
&= 1 - \Gamma^q\left(1 - \frac{\mathrm{R}_0 + \mathrm{R}_l}{\mathrm{R}_c + \mathrm{R}_0}\right) = 1 - \Gamma^q \frac{1 - r_l}{1 + r_0} \\
&= 1 - \exp\left(\operatorname{Log} \Gamma^q \, \frac{1 - r_l}{1 + r_0}\right) \\
&= 1 - \exp\left(-\left[q \operatorname{Log} \frac{1}{\Gamma} + \operatorname{Log} \frac{1 + r_0}{1 - r_l}\right]\right)
\end{aligned}
$$

To compare $\mathrm{I_L}(2q + 1)$ and $\mathrm{I_S}(2q + 1)$ it suffices to compare

$$
q \operatorname{Log} \frac{1}{\Gamma} + \operatorname{Log} \frac{1 + r_0}{1 - r_l}
$$

to $(2q + 1)(r_0 + r_l)$. The first expression is equal to

$$
\begin{aligned}
&q \operatorname{Log} \frac{1}{\Gamma} + \operatorname{Log}(1 + r_0) - \operatorname{Log}(1 - r_l) \\
&\quad \approx 2q\left[r_0 + r_l + \frac{1}{3}(r_0^3 + r_l^3)\right] + \left(r_0 - \frac{r_0^2}{2} + \frac{r_0^3}{3}\right) - \left(-r_l - \frac{r_l^2}{2} - \frac{r_l^3}{3}\right) \\
&\quad = (2q + 1)(r_0 + r_l) + \frac{2q + 1}{3}(r_0^3 + r_l^3) + \frac{r_l^2 - r_0^2}{2} \\
&\quad = (2q + 1)(r_0 + r_l) + (r_0 + r_l)\left[\frac{2q + 1}{3}(r_0^2 - r_0 r_l + r_l^2) - \frac{(r_0 - r_l)}{2}\right]
\end{aligned}
$$

Thus again to a first approximation, if r_0 and r_l are $\ll 1$, $\mathrm{I_L}(2q + 1) - \mathrm{I_S}(2q + 1) \approx 0$, but now the sign of the difference depends on the sign of

$$
\frac{2q + 1}{3}(r_0^2 - r_0 r_l + r_l^2) - \frac{r_0 - r_l}{2}
$$

In terms of q, this is to say that

$$
\begin{aligned}
&\left(2q + 1 < \frac{3}{2} \frac{r_0 - r_l}{r_0^2 - r_0 r_l + r_l^2}\right) \Rightarrow (\mathrm{I_L}(2q + 1) < \mathrm{I_S}(2q + 1)) \\
&\left(2q + 1 > \frac{3}{2} \frac{r_0 - r_l}{r_0^2 - r_0 r_l + r_l^2}\right) \Rightarrow (\mathrm{I_L}(2q + 1) > \mathrm{I_S}(2q + 1))
\end{aligned}
\tag{IV.42}
$$

6.5 *Summary of results* (Figs. 138, 139)

To the extent that $R_0/R_c \ll 1$ and $R_l/R_c \ll 1$, we have $I_L(q) = I_S(q)$. That is, the currents at the middle of the line and in the equivalent inductance are approximately equal for all multiples of the propagation time τ of the line. We have compared the current in the inductance to the current at the middle of the line, because the latter is equal to the input current at

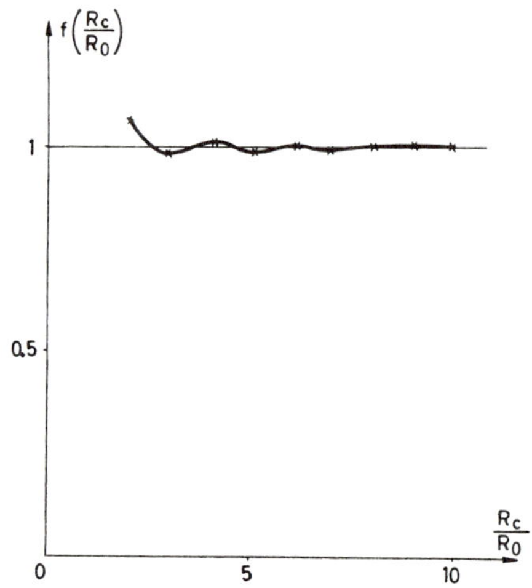

Fig. 138. Ratio of current at the midpoint of the line of Fig. 137 to the current in the equivalent inductance, after one time constant of the inductive circuit, as a function of the source resistance, assuming a short-circuit load.

times $(2q + 1)\tau$, and to the output current at times $2q\tau$. The main advantage of this simplification of the calculations is that the time after which the current is established can easily be calculated. That time is of the order of three time constants of the equivalent inductive circuit, which is to say

$$3 \frac{L_{eq}}{R_0 + R_l} = 3\tau \frac{R_c}{R_0 + R_l}$$

In practice, in problems involving current transfer the load R_l is usually a short circuit, so that $R_l/R_c \ll 1$ is satisfied.

NUMERICAL EXAMPLE. Figure 139 shows $I_L(t/\tau)$ and $I_S(t/\tau)$ for $R_c/R_0=5$ and $R_l = 0$. From the figure, it is seen that, for $t/\tau = 2q$ for any q, $I_L \approx I_S$ but $I_L > I_S$. For $t/\tau = 2q + 1 \leq 7$, $I_L \approx I_S$ but $I_L < I_S$, while for $t/\tau = 2q + 1 \geq 9$, $I_L \approx I_S$ but $I_L > I_S$. This is all in accord with relations (IV.41) and (IV.42) above.

It is interesting also to compare the current in the inductance at the end of one time constant $L_{eq}/R_0 = \tau R_c/R_0$, i.e., $I_S = 1 - e^{-1}$, to the line current at the same instant, for different values of R_c/R_0. Figure 138 shows a plot of

$$\left(\frac{I_L}{I_S}\right)_{t/\tau=R_c/R_0} = f\left(\frac{R_c}{R_0}\right)$$

From this graph it is clear that the ratio of the two currents rapidly tends to unity.

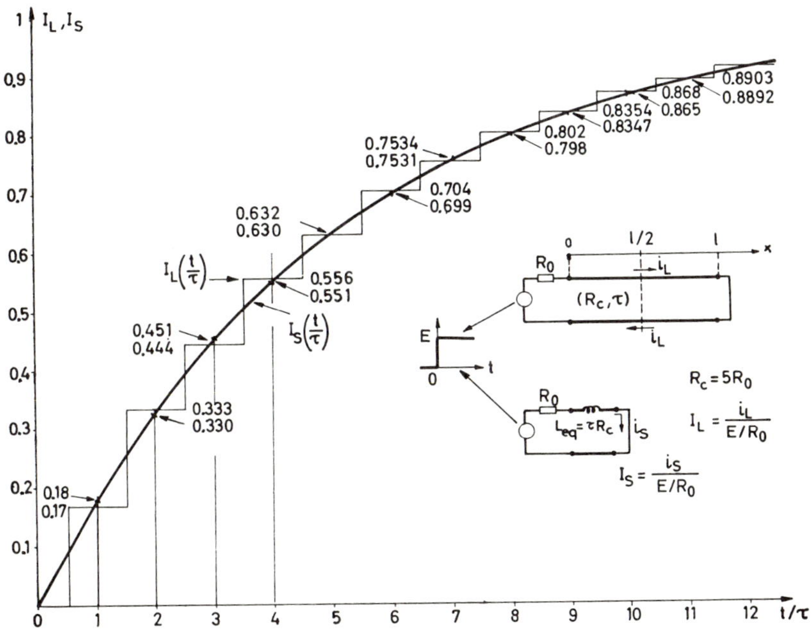

Fig. 139. Current I_L at the midpoint of the line of Fig. 137, and current I_S in the equivalent inductance, for the case $R_0 = R_c/5$ and $R_l = 0$.

It is thus concluded that a line can be replaced by an inductance $L_{eq} = \tau R_c$ whenever it is loaded by resistances $R_0 < R_c$ and $R_l < R_c$, where the resistances are placed as in Fig. 139. (The reflection coefficients Γ_0 and Γ_l are then both negative.)

6.6 *Replacement of the line by a capacitance* (Fig. 140)

The preceding method also can be applied if the input and output reflection coefficients Γ_0 and Γ_l are both positive. This is the case if $R_0 > R_c$ and $R_l > R_c$. It is possible to find an approximation for the evolution of the voltage (not current) at the terminals of the loads. The equivalent capacitance is $C_{eq} = \tau/R_c$. Further, if $R_0 \gg R_c$ and $R_l \gg R_c$, the voltage across C_{eq} is equal to the voltage at the middle of the line ($x = l/2$), for all times $k\tau$ which are multiples of the propagation time τ of the line.

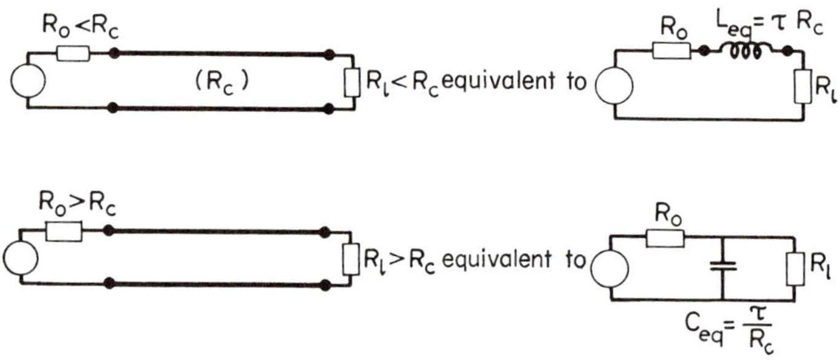

Fig. 140. Equivalent circuits possible in two cases of line loading.

6.7 *The case that Γ_0 and Γ_l are of opposite sign*

If neither the current nor the voltage increases monotonically, the above method can no longer be applied. This is to say that a line loaded by R_0 and R_l is not equivalent to a first-order system if either $R_0 > R_c$ and $R_l < R_c$, or $R_0 < R_c$ and $R_l > R_c$.

7. Appendix

There appear below calculations of the skin effect in both the transient regime and the sinusoidal steady state. The formulas which result for the two regimes are similar, since in the transient regime we introduce an operational depth

$$\delta(p) = \left(\frac{2}{\mu\sigma p}\right)^{1/2}$$

while in the steady state we consider a real depth $(2/\mu\sigma\omega)^{1/2}$.

Consider a metal plate of conductivity σ, of large planar extent, and very thick, traversed by a current I (Fig. 141). The current density has only a component parallel to Oy, call it J(x). Applying Maxwell's equations to this situation† results in

$$\mathbf{rot\,E} = -\frac{\partial \mathbf{B}}{\partial t} = -\mu\frac{\partial \mathbf{H}}{\partial t}$$

$$\mathbf{rot\,H} = \mathbf{J} + \varepsilon\frac{\partial \mathbf{E}}{\partial t} = \sigma\mathbf{E} + \varepsilon\frac{\partial \mathbf{E}}{\partial t}$$

Here the MKS system of units is used, and the various quantities are: **E**, the electric field vector; **B**, the magnetic flux density vector; **H**, the magnetic field vector; **J**, the current density vector (intensity per unit surface traversed).

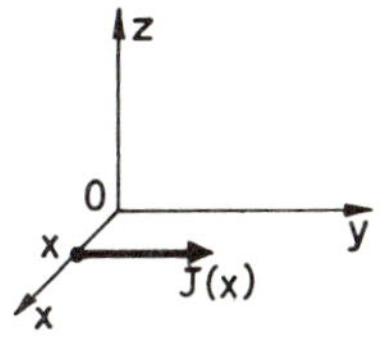

Fig. 141. Coordinates and current density for skin-effect calculations.

Since we are dealing with a conductor, the displacement current is negligible with respect to the conduction current. That is, $\varepsilon(\partial\mathbf{E}/\partial t)$ is much less than $\sigma\mathbf{E}$, since σ is large. There remain

$$\mathbf{rot\,E} = -\mu\frac{\partial \mathbf{H}}{\partial t}$$

$$\mathbf{rot\,H} = \sigma\mathbf{E}$$

7.1 *The transient regime*

Applying the Laplace transformation to the above equations, assuming zero initial conditions and considering **E**, **H**, and **J** to be functions of the complex frequency p, yields

$$\mathbf{rot\,E} = -p\mu\mathbf{H}$$

$$\mathbf{rot\,H} = \sigma\mathbf{E}$$

From these,

$$\mathbf{rot\,rot\,E} = -p\mu\,\mathbf{rot\,H} = -p\mu\sigma\mathbf{E}$$

which is to say

$$\mathbf{grad}\,\mathrm{div}\,\mathbf{E} - \Delta\mathbf{E} = -p\mu\sigma\mathbf{E}$$

† See, e.g., S. Ramo and J. R. Whinnery, "Fields and Waves in Modern Radio." Wiley, New York, 1953.

But

$$\operatorname{div} \mathbf{E} = 0; \qquad \mathbf{grad} \operatorname{div} \mathbf{E} = 0$$

so that there remains only

$$\Delta \mathbf{E} - p\mu\sigma \mathbf{E} = 0$$

Similarly

$$\Delta \mathbf{J} - p\mu\sigma \mathbf{J} = 0$$

Consider now an infinitely thick plane conductor. **J** has only a component $\mathrm{J}(x)$ parallel to $\mathrm{O}y$ and is a function only of the distance x from the surface of the conductor. The equation for J then becomes

$$\frac{d^2\mathrm{J}(x, p)}{dx^2} - p\mu\sigma\, \mathrm{J}(x, p) = 0$$

with solutions of the form

$$\mathrm{J}(x, p) = \mathrm{A}(p) \exp[-(\mu\sigma p)^{1/2}x] + \mathrm{B}(p) \exp[+(\mu\sigma p)^{1/2}x]$$

The function B is identically zero, since the current density can not increase indefinitely with x. The other function is seen to be the density at the surface of the conductor,

$$\mathrm{A} = \mathrm{J}_0(p)$$

so that

$$\mathrm{J}(x, p) = \mathrm{J}_0(p) \exp[-(\mu\sigma p)^{1/2}x]$$

Let us calculate the current which flows in a segment of the conductor of width λ (see Fig. 142). This is

$$\mathrm{I} = \mathrm{J}_0 \int_0^{+\infty} \exp[-(\mu\sigma p)^{1/2}x]\lambda\, dx$$

$$\mathrm{I}(p) = \frac{\lambda}{(\mu\sigma p)^{1/2}} \mathrm{J}_0(p)$$

This last relation shows that the current could be considered to be of uniform density $\mathrm{J}_0(p)$, but flowing only in a skin of thickness

$$\delta(p) = \frac{1}{(\mu\sigma p)^{1/2}}$$

Equally well, any uniform density $\mathrm{J}(p)$ could be assumed in a skin of thickness

$$\delta'(p) = \frac{1}{(\mu\sigma p)^{1/2}} \frac{\mathrm{J}_0(p)}{\mathrm{J}(p)}$$

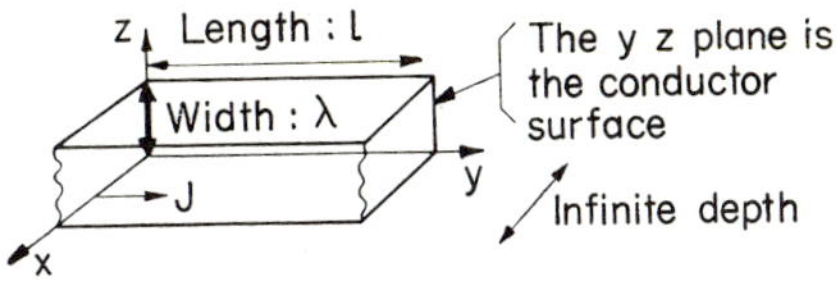

Fig. 142. A conductor segment.

This result could be obtained directly from Ohm's law, expressed in operational notation. Joule's law can not, however, be used in the transient case, as it is below in the sinusoidal steady state, because it is nonlinear. In effect, there is no operational "power," since even if the operator $R(p)$ is constant,

$$W(p) \neq R[I(p)]^2$$

since

$$R[I(p)]^2 = R\left[\int_0^{+\infty} e^{-pt}\, i(t)\, dt\right]^2$$

$$\neq \int_0^{+\infty} e^{-pt}[Ri^2(t)]\, dt$$

Let us consider then the voltage drop across an element of length l, width λ, and of infinite depth (Fig. 143a). The ohmic voltage drop between two planes separated by a distance l will be

$$V(p) = R(p)\, l\, I(p)$$

where, as in the rest of the book, $R(p)$ is the (operational) resistance per

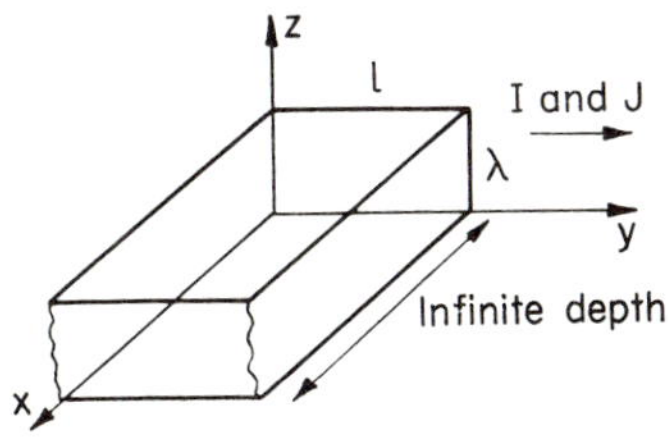

Fig. 143a. Conductor segment.

unit length. $R(p)$ can be calculated as the resistance of a conductor traversed by the same current $I(p)$ distributed uniformly with density $J(p)$ in an area $\lambda\delta(p)$.

In the time domain

$$v(t) = \frac{1}{\sigma}\,\frac{l}{\lambda d}\,i(t)$$

$$= \left(\frac{1}{\sigma}\,\frac{l}{\lambda d}\right)(\lambda d\; j(t)) = \frac{1}{\sigma}\, l\, j(t)$$

Taking the transform of this,

$$\mathrm{V}(p) = \frac{l}{\sigma}\,\mathrm{J}(p)$$

$$= \frac{l}{\sigma}\,\frac{\mathrm{J}(p)}{\mathrm{J}_0(p)}\,\mathrm{J}_0(p)$$

$$= \frac{l}{\sigma}\,\frac{\mathrm{J}(p)}{\mathrm{J}_0(p)}\,\frac{(\mu\sigma p)^{1/2}}{\lambda}\,\mathrm{I}(p)$$

Thus

$$\mathrm{R}(p) = \frac{1}{\sigma}\,\frac{1}{\lambda}\,(\mu\sigma p)^{1/2}\,\frac{\mathrm{J}(p)}{\mathrm{J}_0(p)} = \frac{1}{\sigma}\,\frac{1}{\lambda\,\delta(p)}$$

where

$$\delta(p) = \frac{1}{(\mu\sigma p)^{1/2}}\,\frac{\mathrm{J}_0(p)}{\mathrm{J}(p)}$$

To each value of $\mathrm{J}(p)/\mathrm{J}_0(p)$ there corresponds a different operational skin depth $\delta(p)$. This is an operator relating the transform current $\mathrm{I}(p)$ and the transform current density $\mathrm{J}(p)$, and not the transform of some actual skin depth d.

Let us consider some particular cases.

(a) If $\mathrm{J}_0/\mathrm{J} = 2^{1/2}$, i.e., if the average current density $\mathrm{J}_0/2^{1/2}$ is used, the depth is

$$\delta = (2/\mu\sigma p)^{1/2}$$

in analogy to the result obtained for the sinusoidal case below.

(b) If $\mathrm{J} = \mathrm{J}_0$, the result is

$$\delta = 1/(\mu\sigma p)^{1/2}$$

In the first case,

$$\mathrm{R}(p) = \frac{1}{\sigma}\,\frac{1}{\lambda}\left(\frac{\mu\sigma p}{2}\right)^{1/2} = \mathrm{K}p^{1/2}$$

This formula can be shown to be valid for conductors of various shapes, and is the result used above (Section 5). The value of K can be found experimentally.

7.2 *The sinusoidal steady state*

Writing Maxwell's equations in phasor notation, neglecting the displacement current as above, and considering **E**, **H**, and **J** to be functions of the radian frequency ω, yields

$$\mathbf{rot\ E} = -j\omega\mu\mathbf{H}$$

$$\mathbf{rot\ H} = \sigma\mathbf{E}$$

from which

$$\mathbf{rot\ rot\ E} = -j\omega\mu\ \mathbf{rot\ H} = -j\omega\mu\sigma\mathbf{E}$$

which is to say

$$\mathbf{grad}\ \mathrm{div}\ \mathbf{E} - \Delta\mathbf{E} = -j\omega\mu\sigma\mathbf{E}$$

But

$$\mathrm{div}\ \mathbf{E} = 0; \qquad \mathbf{grad}\ \mathrm{div}\ \mathbf{E} = 0$$

so that there remains only

$$\Delta\mathbf{E} - j\omega\mu\sigma\mathbf{E} = 0$$

Similarly

$$\Delta\mathbf{J} - j\omega\mu\sigma\mathbf{J} = 0$$

Let us consider again an infinitely thick plane conductor. Again J has only a component $\mathrm{J}(x)$ parallel to Oy and is a function only of the distance x below the surface of the conductor. The equation for J then becomes

$$\frac{d^2\mathrm{J}(x, \omega)}{dx^2} - j\omega\mu\sigma\ \mathrm{J}(x, \omega) = 0$$

with physically possible solutions of the form

$$\begin{aligned}\mathrm{J}(x, \omega) &= \mathrm{J}_0 \exp[-j^{1/2}(\mu\sigma\omega)^{1/2}x] \\ &= \mathrm{J}_0 \exp\left[-(1+j)\left(\frac{\mu\sigma\omega}{2}\right)^{1/2} x\right]\end{aligned}$$

where J_0 is the density at the surface of the conductor. We use here the peak value of the sinusoid, and not the rms value.

Let us find the current $\mathrm{I}(\omega)$ flowing in a segment of the conductor (see Fig. 142). This is

$$\mathrm{I} = \mathrm{J}_0 \int_0^{+\infty} \exp[-j^{1/2}(\mu\sigma\omega)^{1/2}x]\lambda\, dx = \frac{\mathrm{J}_0\lambda}{j^{1/2}(\mu\sigma\omega)^{1/2}}$$

Further

$$\frac{1}{j^{1/2}} = \frac{1}{(2^{1/2}/2)(1+j)} = \frac{2^{1/2}(1-j)}{(1+j)(1-j)}$$
$$= (2^{1/2}/2)(1-j)$$

so that

$$\mathrm{I} = \frac{\mathrm{J}_0\lambda}{(2\mu\sigma\omega)^{1/2}}(1-j)$$

and

$$|\,\mathrm{I}\,| = \frac{|\,\mathrm{J}_0\,|\lambda}{(2\mu\sigma\omega)^{1/2}}\,2^{1/2} = \frac{|\,\mathrm{J}_0\,|\lambda}{(\mu\sigma\omega)^{1/2}}$$

in analogy to the expression found above for the transient case.

Remark. The peak values of I and J_0 are to be used here. The effective, or rms, value of I is

$$(\mathrm{I}_{\mathrm{eff}})^2 = \frac{1}{2}(|\,\mathrm{I}\,|)^2 = \frac{1}{2}\frac{\mathrm{J}_0{}^2\lambda^2}{\mu\sigma\omega}$$

By definition

$$\frac{\mathrm{J}_0{}^2}{2} = (\mathrm{J}_{0\ \mathrm{eff}})^2$$

Thus

$$\mathrm{I}_{\mathrm{eff}} = \frac{\mathrm{J}_{0\ \mathrm{eff}}\lambda}{(\mu\sigma\omega)^{1/2}}$$

which again resembles an expression found above for the transient case.

Let us now consider the conductor resistance and the skin depth. In determining this latter, we will use the dissipated power, and hence the effective value of the current.

The power dissipated in a volume of depth dx, width $\lambda = 1$, and length $l = 1$ (Fig. 143b) is

$$\frac{(\mathrm{J}_{\mathrm{eff}})^2}{\sigma}\,\lambda l\,dx = \frac{(\mathrm{J}_{\mathrm{eff}})^2}{\sigma}\,dx$$

so that in the entire (infinite) depth of the plate the power dissipated is

$$\mathrm{W} = \int_0^{+\infty} \frac{(\mathrm{J}_{\mathrm{eff}})^2}{\sigma}\,dx = \frac{1}{\sigma}\int_0^{+\infty} \frac{\mathrm{JJ}^*}{2}\,dx$$

But

$$\mathrm{JJ}^* = \mathrm{J}_0^2 \exp[-xj^{1/2}(\mu\sigma\omega)^{1/2}] \exp[-x(-j)^{1/2}(\mu\sigma\omega)^{1/2}]$$
$$= \mathrm{J}_0^2 \exp[-x(\mu\sigma\omega)^{1/2}(j^{1/2} + (-j)^{1/2})]$$

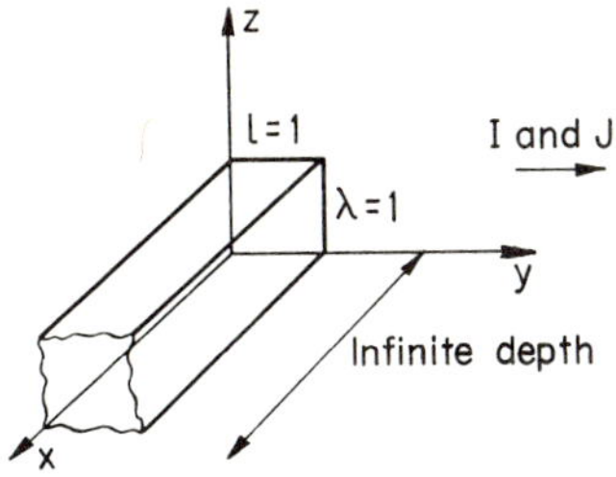

Fig. 143b. Conductor segment.

and since

$$j^{1/2} + (-j)^{1/2} = \frac{1+j}{2^{1/2}} + \frac{1-j}{2^{1/2}} = \frac{2}{2^{1/2}} = 2^{1/2}$$

we have

$$\begin{aligned} JJ^* &= (J_0)^2 \exp[-x(2\mu\sigma\omega)^{1/2}] \\ &= 2(J_{0\,\mathrm{eff}})^2 \exp[-x(2\mu\sigma\omega)^{1/2}] \end{aligned}$$

Thus

$$W = \frac{(J_{0\,\mathrm{eff}})^2}{\sigma} \int_0^{+\infty} \exp[-x(2\mu\sigma\omega)^{1/2}]\, dx = \frac{(J_{0\,\mathrm{eff}})^2}{\sigma(2\mu\sigma\omega)^{1/2}}$$

Let us suppose that the total current is uniformly distributed in a plate of depth d. The power loss in a volume of depth d, width $\lambda = 1$, and length $l = 1$, is

$$RI_{\mathrm{eff}}^2 = \frac{1}{\sigma}\,\frac{1}{d1}\, I_{\mathrm{eff}}^2 = \frac{1}{\sigma d}\,\frac{(J_{0\,\mathrm{eff}})^2 1}{\mu\sigma\omega}$$

To calculate d, we now equate W and RI_{eff}^2:

$$\frac{(J_{0\,\mathrm{eff}})^2}{\sigma(2\mu\sigma\omega)^{1/2}} = \frac{1}{\sigma d}\,\frac{(J_{0\,\mathrm{eff}})^2}{\mu\sigma\omega}$$

Thus

$$d = (2/\mu\sigma\omega)^{1/2}$$

This is a real depth, and not an operational relation.

Let us calculate the effective current density $\mathscr{J}_{\mathrm{eff}}$ across the depth d. (This is of the nature of an equivalent average density in a section of width $\lambda = 1$ and depth d.) See Fig. 144.

$\mathscr{J}_{\mathrm{eff}}$ is defined by

$$\mathscr{J}_{\mathrm{eff}} d1 = I_{\mathrm{eff}}$$

so that

$$\mathscr{J}_{\mathrm{eff}} = \frac{I_{\mathrm{eff}}}{d} = \frac{J_{0\,\mathrm{eff}}}{d(\mu\sigma\omega)^{1/2}} = \frac{J_{0\,\mathrm{eff}}}{2^{1/2}}$$

Thus the current I_{eff} can be considered to be uniformly distributed with density $\mathscr{J}_{eff} = J_{0\,eff}/2^{1/2}$ to a depth

$$d = (2/\mu\sigma\omega)^{1/2}$$

This should be compared to the expressions $J = J_0/2^{1/2}$ and $\delta = (2/\mu\sigma p)^{1/2}$ obtained for the transient case above.

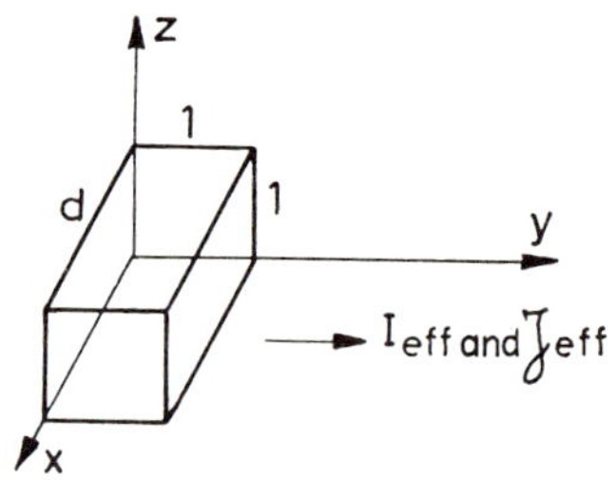

Fig. 144. A conductor segment.

7.3 *Conclusion*

The calculations for the sinusoidal steady state lead to a single unambiguous value d for the skin depth, because two experimental laws are available: Ohm's law and Joule's law. For variable frequency p, only Ohm's law is valid, for only it is linear. The calculations for the transient case then lead only to the form of the skin depth $\delta(p)$, and leave the solution determined only to within a multiplicative constant. In this case, it suffices to replace $j\omega$ by p in the impedances and transfer functions obtained in the sinusoidal case.†

Exercises

The solutions to those exercises noted* will be found at the end of the book, in Chapter VI.

*Exercise 1. Consider a lossless line fed by a voltage generator $E\Upsilon(t)$ with zero internal resistance, terminated in a pure resistance equal to R_c. Find the voltage waveforms at the input to the line, at an arbitrary point, and at the output.

† See A. Angot, "Compléments de Mathématiques," Chapter 8. *Revue d'optique, Paris,* 1957; M. Y. Bernard, "Cours de Radio-électricité générale, année B, du Conservatoire National des Arts et Métiers." Riber, Paris.

*Exercise 2. Consider a lossless line fed by a voltage generator $E\Upsilon(t)$ with internal resistance R_c, terminated in an open circuit. Find the voltage and current at an arbitrary point of the line, and, in particular, at the input and output.

*Exercise 3. The same as Exercise 2, except that the line is terminated in a short circuit.

*Exercise 4. Show that the results of Exercises 1 and 3 can be generalized to the case of a voltage generator of arbitrary emf $e(t)$.

*Exercise 5. Find the output current of a lossless line, having the output short-circuited, and fed by a generator of internal resistance $R_0 < R_c$ and emf a linear ramp terminating at some time T. As a special case, consider $R_c = 50\ \Omega$, $R_0 = 10\ \Omega$, $T = 20$ nsec, and $\tau = 4$ nsec.

*Exercise 6. Consider a lossless line terminated in the series combination of an inductance L_1 and a resistance R_1, fed by a generator with emf $E\Upsilon(t)$ and internal resistance R_0 (Fig. 145).

Fig. 145. Circuit for Exercise 6.

(a) Find $v(l, t)$ and $i(l, t)$ in the case that $R_0 = R_c$, for arbitrary R_1.
(b) Find $v(l, t)$ in the form of a series, in the case that $R_1 = R_c$, for arbitrary R_0.

*Exercise 7. Study the beginning of the transient regime for a line having series loss, excited by a current step $I_0\Upsilon(t)$. The attenuation is sufficiently large that the line may be considered to be semi-infinite (matched at its output).

*Exercise 8. A type of line having low characteristic impedance is the "strip" line of Fig. 146. Let a and b be the width and thickness of the conductors respectively, and let e be the thickness of the dielectric.

(a) Find R_c and α under the assumption that a/e is much larger than unity. Assume that the dielectric is Teflon ($\varepsilon_r = 2.25$) of thickness $e = 0.1$ mm, and that the conductors are copper ($\varrho = 1.7 \times 10^{-8}\ \Omega$meter).
(b) Find the dimensions of lines having $R_c = 50$, 10, and $5\ \Omega$, if an

attenuation of 5% is required over a length of line $l = 20$ cm. Verify the approximation made in part (a) of the question.

(c) Find the skin-effect time constant $T_{0.5}$, following the calculations of Section 7.

(d) Assume that the line is fed by a step with negligible rise time. It is desired that the rise time of the pulse at the output of the above line of length 20 cm be not greater than 1 nsec. Is this possible?

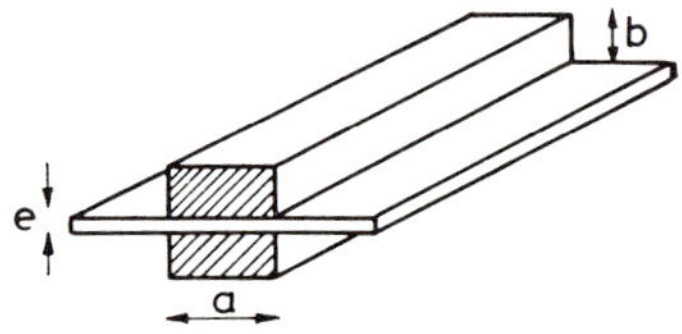

Fig. 146. Strip line studied in Exercise 8.

*EXERCISE 9. Direct current is to be supplied to a variable load through a line having characteristic resistance R_c (this is the problem of supplying power to the components of an apparatus with dimensions large with respect to a wavelength of the signals involved). See Fig. 147. At some instant, a demand for an additional current ΔI is made by the load. This can be represented by the closing of the switch in Fig. 147.

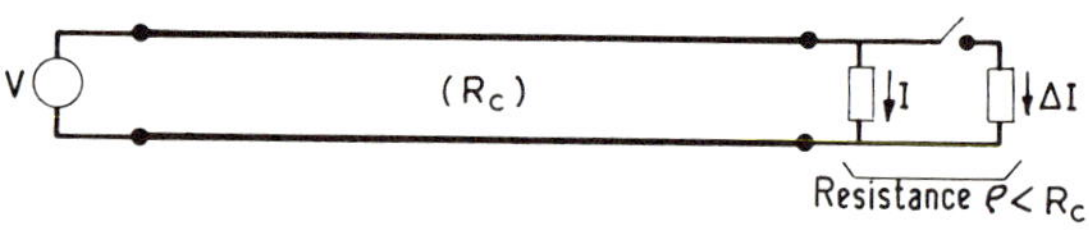

Fig. 147. Circuit for Exercise 9.

(a) Find a lumped-constant equivalent circuit, given that a maximum load voltage change of ΔV is tolerable.

(b) Show that adding a capacitance C in parallel with the load aids in furnishing the extra current demanded ΔI.

(c) Find the best compromise value of the shunt capacitance, assuring small response time and minimum current oscillations in the load.

CHAPTER V

Lines in the Sinusoidal Steady State†

1. Preliminary Definitions

1.1 *The voltage at a point*

The line voltage at an abscissa x in Fig. 148 is

$$v = v_{\mathrm{M}} - v_{\mathrm{N}}$$

in both magnitude and sign. The phase can thus be determined unambiguously with respect to an arbitrary origin. In complex (phasor) notation,

$$\mathrm{V} = |\mathrm{V}|\, e^{j\theta} = |\mathrm{V}| \underline{/\theta}$$

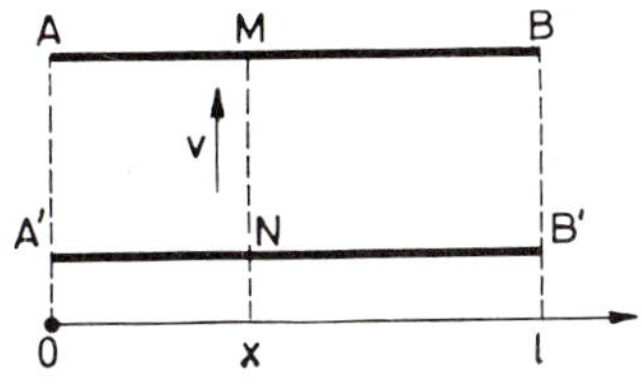

Fig. 148. Voltage convention used.

1.2 *The current at a point*

The current at a point x (Fig. 149) is the forward current from A to B, in magnitude and sign. Its phase can be fixed using as origin the phase

† In this chapter, the sinusoidal steady state will be studied in analogy to the developments above, carried out for pulse operation.

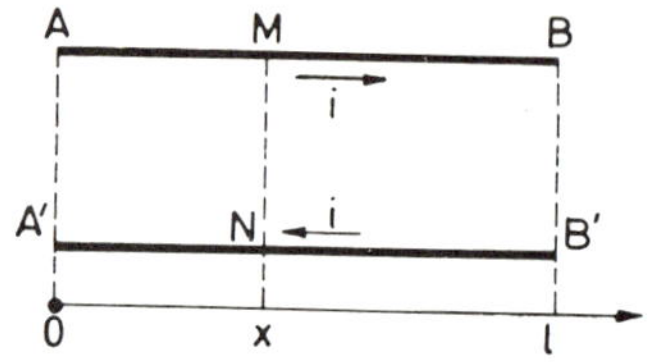

Fig. 149. Current convention used.

origin chosen for the voltage phasor. In complex notation,

$$\mathrm{I} = |\mathrm{I}|\, e^{j\theta'} = |\mathrm{I}|\, \underline{|\theta'}$$

1.3 *Phase of the current with respect to the voltage*

It is often convenient to consider the phase of the current with respect to that of the voltage at the same point (Fig. 150). We will define this phase angle to be

$$\varphi = -(\theta' - \theta)$$

that is, equal in magnitude and opposite in sign to the algebraic phase of the current with respect to the voltage. This results in a phase lag (clockwise rotation) of the current with respect to the voltage being taken positive, contrary to the usual sign convention in trigonometry for measuring angles.

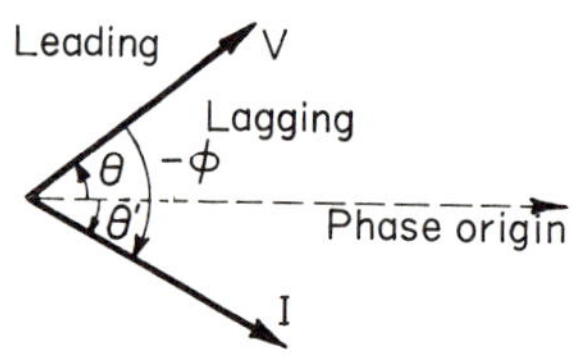

Fig. 150. The current phase is taken positive if it lags the voltage.

Thus a current which is delayed in time with respect to the voltage has a positive phase with respect to the voltage, and vice versa. Thus,

$$\mathrm{I} = |\mathrm{I}|\, \underline{|\theta + \theta' - \theta} = |\mathrm{I}|\, \underline{|\theta - \varphi}$$

If the voltage at point M is taken as defining the phase origin, then we have simply

$$\mathrm{V} = |\mathrm{V}|\,; \qquad \mathrm{I} = |\mathrm{I}|\, \underline{|-\varphi}$$

(ϕ can have any value from 0 to 2π rad.)

2. Solution of the Line Equations, and the Secondary Parameters Z_c and γ

Calculations identical to those of Section 5 of Chapter I can be carried out for the sinusoidal steady state. Since the electromotive force applied to the network is sinusoidal, so also are the voltage and current at every point of the network. Thus the complex number (phasor) notation can be used to advantage. Thus the current $i(x, t)$ will be taken as the real part of the complex function $\mathscr{I}(x) \exp(j\omega t)$, and the voltage as the real part of $\mathscr{V}(x) \exp(j\omega t)$. The time dependence will then be suppressed in the usual way.

Equations (I.3) and (I.4) then become

$$\frac{\partial \mathscr{V}}{\partial x} = -(\mathrm{R} + j\mathrm{L}\omega)\mathscr{I}$$
$$\frac{\partial \mathscr{I}}{\partial x} = -(\mathrm{G} + j\mathrm{C}\omega)\mathscr{V} \qquad \text{(V.1)}$$

from which

$$\frac{\partial^2 \mathscr{V}}{\partial x^2} = (\mathrm{R} + j\mathrm{L}\omega)(\mathrm{G} + j\mathrm{C}\omega)\mathscr{V}$$
$$\frac{\partial^2 \mathscr{I}}{\partial x^2} = (\mathrm{R} + j\mathrm{L}\omega)(\mathrm{G} + j\mathrm{C}\omega)\mathscr{I}$$

Defining the admittance per unit length

$$\mathrm{Y}(\omega) = \mathrm{G} + j\mathrm{C}\omega$$

and impedance per unit length

$$\mathrm{Z}(\omega) = \mathrm{R} + j\mathrm{L}\omega$$

as well as

$$\gamma^2 = (\mathrm{R} + j\mathrm{L}\omega)(\mathrm{G} + j\mathrm{C}\omega)$$

we have

$$\frac{\partial^2 \mathscr{V}}{\partial x^2} = \gamma^2 \mathscr{V}$$

with solution

$$\mathscr{V}(x) = \mathrm{V}_1 e^{-\gamma x} + \mathrm{V}_2 e^{\gamma x}$$

where V_1 and V_2 are complex functions of $j\omega$, but independent of x.

The current can be obtained by differentiating the voltage:

$$\frac{\partial \mathscr{V}}{\partial x} = [-\mathrm{V}_1 e^{-\gamma x} + \mathrm{V}_2 e^{\gamma x}]\gamma = -[\mathrm{R} + j\mathrm{L}\omega]\mathscr{I} = -\mathrm{Z}\mathscr{I}$$

from which

$$\mathcal{I}(x) = V_1 \frac{\gamma}{Z} e^{-\gamma x} - V_2 \frac{\gamma}{Z} e^{\gamma x}$$

Defining

$$\frac{\gamma}{Z} = \left(\frac{Y}{Z}\right)^{1/2} = \frac{1}{Z_c}$$

this becomes

$$\mathcal{I}(x) = \frac{V_1}{Z_c} e^{-\gamma x} - \frac{V_2}{Z_c} e^{\gamma x}$$

The solution of the system (V.1) is thus

$$\begin{aligned} \mathcal{V}(x) &= V_1 e^{-\gamma x} + V_2 e^{\gamma x} \\ \mathcal{I}(x) &= \frac{V_1}{Z_c} e^{-\gamma x} - \frac{V_2}{Z_c} e^{\gamma x} \end{aligned} \qquad \text{(V.2)}$$

with

$$v(x, t) = \mathrm{Re}[\mathcal{V}(x) e^{j\omega t}]$$
$$i(x, t) = \mathrm{Re}[\mathcal{I}(x) e^{j\omega t}]$$

where

$$\gamma = [(R + jL\omega)(G + jC\omega)]^{1/2}$$
$$Z_c = \left(\frac{R + jL\omega}{G + jC\omega}\right)^{1/2}$$

Henceforth we shall write simply

$$\mathcal{V}(x) = V$$
$$\mathcal{I}(x) = I$$

Suppose that the current and voltage at some point A are known, say I_0 and V_0. Let A be taken as the distance origin. Thus for $x = 0$

$$V_0 = V_1 + V_2; \qquad V_1 = \frac{V_0 + Z_c I_0}{2}$$
$$I_0 = \frac{V_1 - V_2}{Z_c}; \qquad V_2 = \frac{V_0 - Z_c I_0}{2}$$

so that

$$V = V_0 \cosh \gamma x - Z_c I_0 \sinh \gamma x$$
$$I = -\frac{V_0}{Z_c} \sinh \gamma x + I_0 \cosh \gamma x$$

These relate the current and voltage at an arbitrary point along the line to the current and voltage at some other point A. Recall that γ and Z_c are complex numbers in these. These equations assume the sign conventions used above for the measurement of current and distance, but are independent of the direction in which energy is actually propagated.

3. Characteristic Impedance for Some Special Cases

We have

$$Z_c = \left(\frac{R + jL\omega}{G + jC\omega}\right)^{1/2} = \left(\frac{L}{C}\right)^{1/2}\left(\frac{R/L\omega + j}{G/C\omega + j}\right)^{1/2}$$

$$= \left(\frac{L}{C}\right)^{1/2}\left(\frac{(R/L\omega + j)(G/C\omega - j)}{(G/C\omega)^2 + 1}\right)^{1/2}$$

$$= \left(\frac{L}{C}\right)^{1/2}\left(\frac{1 - j(R/L\omega - G/C\omega) + RG/LC\omega^2}{(G/C\omega)^2 + 1}\right)^{1/2}$$

3.1 *Lines with small losses*

If $R/L\omega$ and $G/C\omega$ are sufficiently small that their squares and their product are much less than unity (small losses), then

$$Z_c \approx \left(\frac{L}{C}\right)^{1/2}\left(1 - j\frac{R/L\omega - G/C\omega}{2}\right)$$

3.2 *The Heaviside condition is satisfied*

If $R/L = G/C$, then

$$Z_c = (L/C)^{1/2}$$

is a pure resistance. But this condition rarely holds, and then only over a narrow frequency band, because of the frequency variation of R, L, G, and C. This case is thus not of interest if the line is to carry wide-band signals.

3.3 *Lossless lines*

If there are no losses, $R = G = 0$, then $Z_c = (L/C)^{1/2}$ is real and independent of frequency. It is then referred to as the characteristic resistance R_c.

4. Interpretation of γ

4.1 *Preliminary remarks*

γ is the current propagation function (we prefer this terminology to "propagation constant," since γ is not constant, but a function of ω). We have

$$\gamma = \alpha + j\beta = [(\mathrm{R} + j\mathrm{L}\omega)(\mathrm{G} + j\mathrm{C}\omega)]^{1/2}$$

where α is the attenuation function and β is the wavelength function, or the phase function. γ has the dimension of inverse length, and is a per unit length quantity.

4.2 *Signs of α and β*

We will show that α and β are both > 0. The line is a passive system, and can not produce energy. The amplitude of a propagating wave is multiplied by $\exp(-\alpha x)$ (see Section 5), and thus α certainly can not be negative, since this would imply generation of energy along the line.

As to β, it can not be negative, since if that were the case, it would imply that effect preceded cause. It will be shown in Section 5 below that in the (x, t) plane, the curves of equal phase, $\omega t - \beta x = \text{const}$, are parallel lines with positive slope, so that β is positive. If the slope were negative, we would have $t_2 > t_1$ with $x_2 < x_1$, for the same line conditions, which is not possible for an incident wave. The same reasoning can be applied to a reflected wave, to conclude again that β is positive.

5. Decomposition of the Steady State into Two Traveling Waves

5.1 *Incident and reflected waves*

At an arbitrary point of the circuit,

$$\mathrm{V} = \mathrm{V}_1 e^{-\gamma x} + \mathrm{V}_2 e^{\gamma x}$$

$$\mathrm{I} = \frac{\mathrm{V}_1 e^{-\gamma x} - \mathrm{V}_2 e^{\gamma x}}{\mathrm{Z}_c}$$

where V_1, V_2, V_c, and γ are complex numbers. Let us write

$$\mathrm{Z}_c = |\mathrm{Z}_c| \underline{|\zeta}$$

$$\gamma = \alpha + j\beta$$

$$\mathrm{V}_1 = |\mathrm{V}_1| \underline{|\mu}$$

$$\mathrm{V}_2 = |\mathrm{V}_2| \underline{|\nu}$$

so that the above relations become

$$\mathrm{V} = |\mathrm{V}_1|\, e^{-\alpha x}\, \underline{|\ \mu - \beta x} \qquad + |\mathrm{V}_2|\, e^{\alpha x}\, \underline{|\ \nu + \beta x}$$

$$\mathrm{I} = \frac{|\mathrm{V}_1|}{|\mathrm{Z}_c|}\, e^{-\alpha x}\, \underline{|\ \mu - \beta x - \zeta} - \frac{|\mathrm{V}_2|}{|\mathrm{Z}_c|}\, e^{\alpha x}\, \underline{|\ \nu + \beta x - \zeta}$$

The corresponding time functions are

$$v = |\mathrm{V}_1| e^{-\alpha x} \cos(\omega t - \beta x + \mu) + |\mathrm{V}_2| e^{\alpha x} \cos(\omega t + \beta x + \nu)$$

$$i = \frac{|\mathrm{V}_1|}{|\mathrm{Z}_c|} e^{-\alpha x} \cos(\omega t - \beta x + \mu - \zeta) - \frac{|\mathrm{V}_2|}{|\mathrm{Z}_c|} e^{\alpha x} \cos(\omega t + \beta x + \nu - \zeta)$$

Let us define

$$v_i = |\mathrm{V}_1| e^{-\alpha x} \cos(\omega t - \beta x + \mu); \qquad v_r = \quad |\mathrm{V}_2| e^{\alpha x} \cos(\omega t + \beta x + \nu)$$

$$i_i = \frac{|\mathrm{V}_1|}{|\mathrm{Z}_c|} e^{-\alpha x} \cos(\omega t - \beta x + \mu - \zeta);\ i_r = -\frac{|\mathrm{V}_2|}{|\mathrm{Z}_c|} e^{\alpha x} \cos(\omega t + \beta x + \nu - \zeta)$$

so that

$$v = v_i + v_r$$

$$i = i_i + i_r$$

Each pair of functions (v_i, i_i), (v_r, i_r) constitutes a system of traveling waves. The two wave systems propagate in opposite directions with the same speed ω/β, and are exponentially attenuated in the direction of propagation. For example, consider the term v_i. The curves of equal phase

$$\omega t - \beta x + \mu = \mathrm{C}$$

that is

$$\omega t - \beta x = \mathrm{C}'$$

are parallel straight lines in the (x, t) plane, with slope ω/β, since

$$\frac{dx}{dt} = \frac{\omega}{\beta} = u$$

The phase, and thus v_i, has the same value at times t_1, t_2 for abscissas x_1, x_2, respectively, as shown in Fig. 151. Thus v_i is a wave traveling in the positive direction (from A towards B), with constant velocity ω/β. Since the factor $\exp(-\alpha x)$ decreases with increasing x, the wave decreases exponentially in propagating in the direction of increasing x. With the voltage wave v_i there is associated a current wave i_i, of amplitude $|i_i| = |v_i|/|\mathrm{Z}_c|$, which lags the voltage wave in phase by the characteristic phase ζ. The current wave propagates with, and is attenuated along with, the voltage wave.

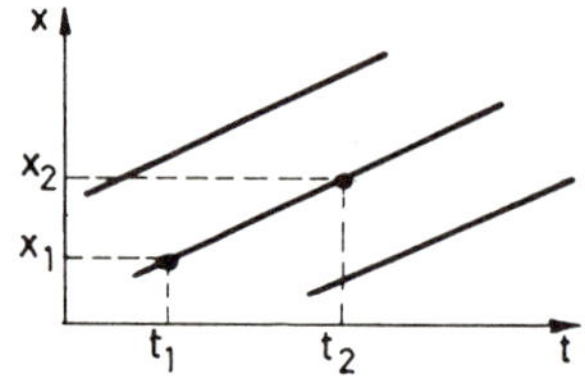

Fig. 151. Lines of constant phase for incident wave.

The same reasoning applied to the function v_r shows that it represents a wave traveling in the negative direction, that is, from B towards A, with the same constant speed ω/β. This wave is also attenuated exponentially, since the factor $\exp(\alpha x)$ decreases when x decreases. To the voltage wave v_r there also corresponds a current wave i_r, which propagates and is attenuated in the same way as the voltage wave. Its amplitude is $|i_r| = |v_r|/|Z_c|$, but it is of opposite sign to the voltage wave, and, in addition, delayed in phase by the characteristic phase ζ.

For the usual case of energy transmission from A towards B, $|v_r| < |v_i|$. v_i is the incident wave, of greater amplitude than v_r, which is the reflected wave.

In summary, knowledge of the characteristic impedance Z_c in both magnitude and phase suffices to characterize the relations between voltage and current for the two systems of waves: $v_i = Z_c i_i$, $v_r = -Z_c i_r$. The propagation characteristics of the two wave systems are entirely determined by the real part α and the imaginary part β of the propagation function γ, which justifies the terminology of Section 4.1: α is a function of the attenuation, and $\beta = 2\pi/\lambda$ is a function of the wavelength. This last follows from

$$\lambda = u\mathrm{T} = \frac{u}{f} = \frac{1}{f}\left(\frac{\omega}{\beta}\right) = \frac{1}{f}\,\frac{2\pi f}{\beta} = \frac{2\pi}{\beta}$$

5.2 *Instantaneous power*

We have the instantaneous power as

$$vi = (v_i + v_r)(i_i + i_r) = v_i i_i + v_r i_r + (v_i i_r + v_r i_i)$$

Let us find the condition under which the cross terms $v_i i_r + v_r i_i$ vanish identically. For this, we must have for all x and t

$$|\mathrm{V}_1|e^{-\alpha x}\left(\frac{-|\mathrm{V}_2|}{|\mathrm{Z}_c|}\,e^{+\alpha x}\right)\cos(\omega t - \beta x + \mu)\cos(\omega t + \beta x + \nu - \zeta)$$

$$+\,|\mathrm{V}_1|e^{+\alpha x}\left(\frac{|\mathrm{V}_2|}{|\mathrm{Z}_c|}\,e^{-\alpha x}\right)\cos(\omega t + \beta x + \nu)\cos(\omega t - \beta x + \mu - \zeta) \equiv 0$$

This requires $\zeta = 0$, and since

$$Z_c = \left(\frac{R + jL\omega}{G + jC\omega}\right)^{1/2}$$

there results finally the familiar Heaviside condition, $R/L = G/C$, which, in particular, holds for a lossless line ($R = G = 0$). For this latter case, the characteristic impedance is a pure resistance $R_c = (L/C)^{1/2}$.

In short, the decomposition into traveling waves is not applicable for power calculations, unless the line satisfies the Heaviside condition. This condition is, however, satisfied for a lossless line, among other cases.

6. The Matched Line

6.1 *Interpretation of* Z_c

Consider a semi-infinite line, as in Fig. 152. The voltage and current along any line are

$$\mathscr{V}(x) = V_1 e^{-\gamma x} + V_2 e^{+\gamma x}$$

$$\mathscr{I}(x) = \frac{V_1}{Z_c} e^{-\gamma x} - \frac{V_2}{Z_c} e^{+\gamma x}$$

Since neither $\mathscr{V}$ nor $\mathscr{I}$ can increase indefinitely with increasing x, for the semi-infinite line it is necessary that $V_2 = 0$. This results in $\mathscr{V}(x)/\mathscr{I}(x)$

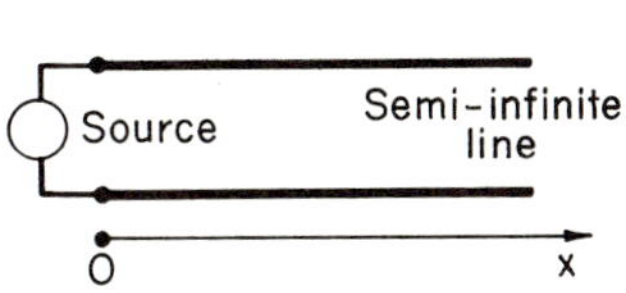

Fig. 152. A semi-infinite line.

$= Z_c =$ const. Z_c is called the characteristic impedance of the line, and is the input impedance of a semi-infinite line. The characteristic impedance is independent of x.

6.2 *Line terminated in its characteristic impedance*

Consider the case of Fig. 153, of a line terminated in its characteristic impedance Z_c. According to Section 2, the output voltage in general is

$$\mathscr{V}(l) = V_1 e^{-\gamma l} + V_2 e^{+\gamma l}$$

$$\mathscr{I}(l) = \frac{V_1}{Z_c} e^{-\gamma l} - \frac{V_2}{Z_c} e^{+\gamma l}$$

The load impedance Z_c imposes the condition

$$\mathscr{V}(l) = Z_c \mathscr{I}(l)$$

from which

$$V_1 e^{-\gamma l} + V_2 e^{+\gamma l} = V_1 e^{-\gamma l} - V_2 e^{+\gamma l}$$

This in turn requires $V_2 = 0$. Thus there is no reflected wave. At each point of the line

$$\mathscr{V}(x) = V_1 e^{-\gamma x}$$

$$\mathscr{I}(x) = \frac{V_1}{Z_c} e^{-\gamma x}$$

That is, for a line terminated in its characteristic impedance, at each point along the line the impedance seen is equal to the characteristic impedance, regardless of the length of the line. This is the characteristic regime, or matched condition, for which no reflected waves are present.

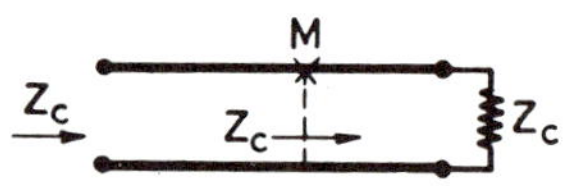

Fig. 153. Line terminated in its characteristic impedance.

Remark. It is not possible to construct a lumped impedance equal to $[(R + jL\omega)/(G + jC\omega)]^{1/2}$, the characteristic impedance, for all frequencies ω, or in fact for more than one frequency, unless $R = G = 0$ (lossless line).

7. Approximate Values of γ. Propagation Velocity as a Function of Frequency (Dispersion)

Recall that

$$\gamma = [(R + jL\omega)(G + jC\omega)]^{1/2}$$

1. *The lossless line*

For $R = G = 0$,

$$\gamma = j\beta = j\omega(LC)^{1/2}; \qquad u = \frac{\omega}{\beta} = \frac{1}{(LC)^{1/2}}; \qquad \lambda = \frac{2\pi}{\beta}$$

The propagation velocity u is independent of frequency.

2. *Line with small losses*

Suppose that

$$\left(\frac{R}{L\omega}\right)^2 \ll 1; \qquad \left(\frac{G}{C\omega}\right)^2 \ll 1; \qquad \left(\frac{RG}{LC\omega^2}\right) \ll 1$$

Then

$$\gamma = \omega(LC)^{1/2}\left[\left(\frac{R}{L\omega} + j\right)\left(\frac{G}{C\omega} + j\right)\right]^{1/2}$$

$$\approx \omega(LC)^{1/2}\left[-1 + j\left(\frac{R}{L\omega} + \frac{G}{C\omega}\right)\right]^{1/2}$$

$$\gamma \approx j\omega(LC)^{1/2}\left[1 - j\left(\frac{R}{L\omega} + \frac{G}{C\omega}\right)\right]^{1/2} \approx \omega(LC)^{1/2}\left(\frac{1}{2}\left(\frac{R}{L\omega} + \frac{G}{C\omega}\right) + j\right)$$

from which

$$\alpha = \frac{\omega(LC)^{1/2}}{2}\left[\frac{R}{L\omega} + \frac{G}{C\omega}\right] = \frac{(LC)^{1/2}}{2}\left(\frac{R}{L} + \frac{G}{C}\right)$$

$$\beta = \omega(LC)^{1/2}$$

and

$$u = \frac{\omega}{\beta} = \frac{1}{(LC)^{1/2}}$$

Here the propagation velocity u is the same as for the lossless line, as is $\lambda = 2\pi/\beta$. There is however an attenuation function α,

$$\alpha = \frac{R}{2R_c} + \frac{GR_c}{2}$$

which is independent of frequency.

3. *General case*

Let p and q be defined by

$$\frac{R}{L\omega} = \sinh p; \qquad \frac{G}{C\omega} = \sinh q$$

It can be shown that

$$\gamma = \omega(LC)^{1/2}\left[\sinh\frac{p+q}{2} + j\cosh\frac{p-q}{2}\right] = \alpha + j\beta$$

from which

$$\beta = \omega(LC)^{1/2}\cosh\frac{p-q}{2}$$

($\omega(LC)^{1/2}$ is the β for a lossless line.) In general, this value of β is larger than that for a lossless line, so that

$$u = \frac{\omega}{\beta} < \frac{\omega}{\beta_{\text{lossless}}} = u_{\text{lossless}}$$

There is a dispersion of propagation velocity with frequency.

4. *The Heaviside condition is satisfied*

If R/L = G/C, then $p = q$, $\beta = \omega(LC)^{1/2}$, and $u = 1/(LC)^{1/2}$.

5. *Application of the above results*

Using the Fourier transform,* a pulse of any shape can be decomposed into the superposition of an infinite number of sinusoids, which constitute its spectrum. These elementary waves propagate independently, each with its proper velocity. If there is no velocity dispersion, the superposition of the elementary waves, each with the same velocity, reconstitutes the original wave, which then propagates with the same velocity $u = 1/(LC)^{1/2}$ as the elementary waves. This is the situation for a lossless line, an idealization useful as the first approximation to a real line, or for a line satisfying the Heaviside condition (a case of no practical interest, since such lines are no longer constructed).

If velocity dispersion is present, the elementary waves propagate with different velocities, with the result that if the elementary waves are superposed at some point along the line, a waveform results which is different from that present at the input of the line. Use was made of this property in Section 5.1 of Chapter II.

8. Impedance Seen Looking into a Line

8.1 *Expression for the impedance seen*

Consider a line loaded by an impedance Z_l, as in Fig. 154. We noted in Section 2 that the current and voltage at time t at a point x could be written in complex notation

$$\mathscr{V}(x)e^{j\omega t}$$

$$\mathscr{I}(x)e^{j\omega t}$$

* See G. Metzger and J.-P. Vabre, "Electronique des Impulsions," Vol. I, "Circuits à Constantes Localisées," Chapter I. Masson, Paris, 1966.

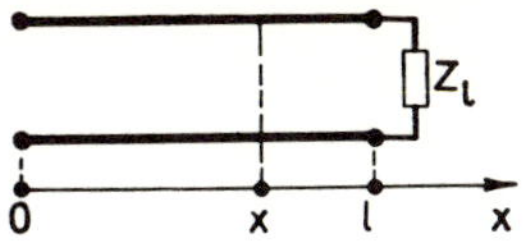

Fig. 154. Line with general load.

Thus the impedance seen looking into the line at abscissa x is

$$Z(x) = \frac{\mathscr{V}(x)e^{j\omega t}}{\mathscr{I}(x)e^{j\omega t}} = \frac{\mathscr{V}(x)}{\mathscr{I}(x)}$$

and is independent of time. Using (V.2) this is

$$Z(x) = Z_c \frac{V_1 e^{-\gamma x} + V_2 e^{\gamma x}}{V_1 e^{-\gamma x} - V_2 e^{\gamma x}} = Z_c \frac{1 + (V_2/V_1)e^{2\gamma x}}{1 - (V_2/V_1)e^{2\gamma x}} \tag{V.3}$$

Using the fact that

$$Z(l) = Z_l$$

from (V.3) we have

$$\frac{Z_l}{Z_c} = \frac{1 + (V_2/V_1)e^{2\gamma l}}{1 - (V_2/V_1)e^{2\gamma l}}$$

from which

$$\frac{V_2}{V_1} e^{2\gamma l}\left(1 + \frac{Z_l}{Z_c}\right) = \frac{Z_l}{Z_c} - 1$$

$$\frac{V_2}{V_1} = \frac{(Z_l/Z_c) - 1}{(Z_l/Z_c) + 1} e^{-2\gamma l}$$

Defining the normalized impedances

$$z_l = \frac{Z_l}{Z_c}; \qquad z(x) = \frac{Z(x)}{Z_c}$$

this becomes

$$\frac{V_2}{V_1} = \frac{z_l - 1}{z_l + 1} e^{-2\gamma l}$$

Substituting into (V.3) yields

$$\frac{Z(x)}{Z_c} = z(x) = \frac{1 + [(z_l - 1)/(z_l + 1)]e^{-2\gamma(l-x)}}{1 - [(z_l - 1)/(z_l + 1)]e^{-2\gamma(l-x)}}$$

or

$$z(x) = \frac{z_l + 1 + (z_l - 1)e^{-2\gamma(l-x)}}{z_l + 1 - (z_l - 1)e^{-2\gamma(l-x)}} = \frac{z_l(1 + e^{-2\gamma(l-x)}) + 1 - e^{-2\gamma(l-x)}}{z_l(1 - e^{-2\gamma(l-x)}) + 1 + e^{-2\gamma(l-x)}}$$

$$z(x) = \frac{z_l + [(1 - e^{-2\gamma(l-x)})/(1 + e^{-2\gamma(l-x)})]}{1 + z_l[(1 - e^{-2\gamma(l-x)})/(1 + e^{-2\gamma(l-x)})]} = \frac{z_l + \tanh \gamma(l - x)}{1 + z_l \tanh \gamma(l - x)}$$

This last is the normalized impedance seen at a point x along a line loaded by a normalized impedance z_l. This expression for $z(x)$ leads to a simple geometric representation, called the circular representation of impedances. We will not take up this subject here, since this diagram and others derived from it are of most use in uhf circuits (transmission lines and waveguides).

8.2 *Physical interpretation*

Let us define a complex number $\psi = \xi + j\eta$ by

$$\tanh \psi = z_l = \frac{Z_l}{Z_c}$$

Then

$$z(x) = \frac{\tanh \psi + \tanh \gamma(l - x)}{1 + \tanh \psi \tanh \gamma(l - x)} = \tanh[\psi + \gamma(l - x)]$$

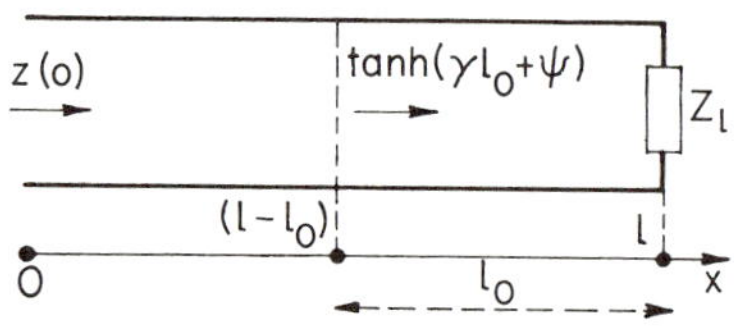

Fig. 155. Normalized impedance seen looking into a line a distance l_0 back from the load.

Let l_0 be the length of a segment of the line measured back towards the source from the load (Fig. 155). The impedance seen looking into the line, at $x = 0$, is

$$\begin{aligned} z(0) &= \tanh(\gamma l + \psi) \\ &= \tanh[\gamma(l - l_0) + \gamma l_0 + \psi] \\ &= \frac{\tanh \gamma(l - l_0) + \tanh(\gamma l_0 + \psi)}{1 + \tanh \gamma(l - l_0) \tanh(\gamma l_0 + \psi)} \end{aligned}$$

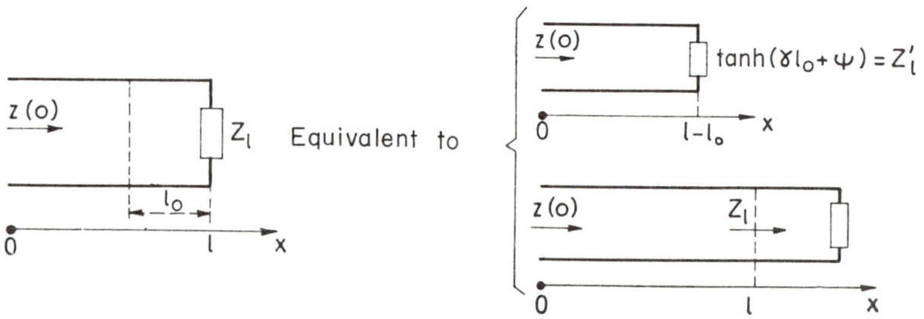

Fig. 156. Replacement of a segment of line by the impedance seen looking into it.

But $\tanh(\gamma l_0 + \psi)$ is the impedance seen looking into the segment of line of length l_0. Hence the impedance seen looking into a line is unchanged if the line is cut at any point, and the line segment and original load removed and replaced by the impedance originally seen looking into the line towards the load at the point where the cut was made (Fig. 156). Also as indicated in Fig. 156, affairs are unchanged if the original load Z_l is replaced by a loaded line segment having input impedance Z_l.

8.3 *The lossless line*

For this case, $\alpha = 0$ so that $\gamma = j\beta$. Then

$$z(x) = \frac{z_l + \tanh \gamma(l - x)}{1 + z_l \tanh \gamma(l - x)} = \frac{z_l + \tanh[j\beta(l - x)]}{1 + z_l \tanh[j\beta(l - x)]}$$

But

$$\tanh(ju) = j \tan u$$

so that

$$z(x) = \frac{z_l + j \tan[\beta(l - x)]}{1 + jz_l \tan[\beta(l - x)]}$$

In particular, the impedance seen looking into the input of a lossless line is

$$z(0) = \frac{z_l + j \tan \beta l}{1 + jz_l \tan \beta l}$$

9. The Lossless Line as an Impedance Transformer

The formulas of Section 8.3 above allow calculation of how the impedance, voltage, and current at the load are transformed when seen through the transmission line. We have already encountered such an impedance transformation in Section 3.7 of Chapter II.

9.1 *Line terminated in its characteristic resistance*

For this case,

$$Z_l = R_c$$

$$z_l = \frac{Z_l}{R_c} = \frac{R_c}{R_c} = 1$$

and

$$z(x) = \frac{1 + j \tan[\beta(l - x)]}{1 + j \tan[\beta(l - x)]} = 1$$

for all x. In particular, $z(0) = 1$, and hence $Z(0) = R_c$. That is, a lossless line terminated in its characteristic resistance presents at its input a pure resistance R_c.

9.2 *Line of length equal to an integral number of wavelengths*

Consider in general moving an integral number of wavelengths along a line. Since

$$\beta\lambda = \frac{2\pi}{\lambda}\,\lambda = 2\pi$$

we have for any integral k, positive, negative, or zero, that

$$\mathscr{V}(x + k\lambda) = V_1 e^{-j\beta(x+k\lambda)} + V_2 e^{j\beta(x+k\lambda)}$$
$$\mathscr{V}(x + k\lambda) = V_1 e^{-j\beta x} + V_2 e^{j\beta x} = \mathscr{V}(x)$$

In the same way it can be seen that

$$\mathscr{I}(x + k\lambda) = \mathscr{I}(x)$$

so that further

$$Z(x + k\lambda) = \frac{\mathscr{V}(x + k\lambda)}{\mathscr{I}(x + k\lambda)} = \frac{\mathscr{V}(x)}{\mathscr{I}(x)} = Z(x)$$

9.3 *The half-wave line*

Consider the transformations resulting from moving in either direction an integral number of wavelengths, plus a half wavelength. Since

$$\beta\lambda = \frac{2\pi}{\lambda}\,\lambda = 2\pi \qquad \text{and} \qquad e^{\pm j\pi} = -1$$

we have

$$\mathscr{V}\left(x + (2k+1)\frac{\lambda}{2}\right) = V_1 \exp\left(-j\beta\left[x + (2k+1)\frac{\lambda}{2}\right]\right)$$
$$+ V_2 \exp\left(+j\beta\left[x + (2k+1)\frac{\lambda}{2}\right]\right)$$
$$= -V_1 e^{-j\beta x} - V_2 e^{j\beta x} = -\mathscr{V}(x)$$

In the same way it can be seen that

$$\mathscr{I}\left(x + (2k+1)\frac{\lambda}{2}\right) = -\mathscr{I}(x)$$

$$Z\left(x + (2k+1)\frac{\lambda}{2}\right) = Z(x)$$

9.4 *The quarter-wave line*

Consider moving in either direction an integral number of half wavelengths, plus a quarter wavelength. Since

$$\beta\lambda = 2\pi$$

$$\exp\left(+ j\frac{\pi}{2}\right) = +j$$

$$\exp\left(- j\frac{\pi}{2}\right) = -j$$

we have

$$\mathscr{V}\left(x + (2k+1)\frac{\lambda}{4}\right) = (-1)^k(-j)[V_1 e^{-j\beta x} - V_2 e^{j\beta x}]$$
$$= (-1)^{k+1} j \mathscr{I}(x)\, R_c$$

and in addition

$$\mathscr{I}\left(x + (2k+1)\frac{\lambda}{4}\right) = (-1)^{k+1} j \frac{\mathscr{V}(x)}{R_c}$$

$$Z\left(x + (2k+1)\frac{\lambda}{4}\right) = \frac{R_c^2}{Z(x)}$$

This last relation shows that a "quarter-wave transformer", i.e., a segment of lossless line of length $(2k+1)\lambda/4$, transforms a terminating impedance Z_l into $Z(0) = R_c^2/Z_l$. In particular, an open-circuited quarter-wave line appears at its input as a short circuit, and vice versa.

9.5 *The eighth-wave line*

Consider a length of line equal to an integral number of quarter wavelengths, plus an eighth wavelength. Since

$$\tan \beta l = \tan\left(\beta(2k+1)\frac{\lambda}{8}\right) = \tan\left((2k+1)\frac{\pi}{4}\right) = \tan\left(k\frac{\pi}{2} + \frac{\pi}{4}\right)$$
$$= \frac{\tan k(\pi/2) + \tan(\pi/4)}{1 - \tan(\pi/4)\tan k(\pi/2)} = \frac{1 + \tan k(\pi/2)}{1 - \tan k(\pi/2)} = (-1)^k$$

we have

$$z(0) = \frac{z_l + (-1)^k j}{1 + (-1)^k j z_l}$$

Thus if z_l is real, $|z(0)| = 1$.

In particular, if the line is short-circuited, $z_l = 0$, then $z(0) = (-1)^k j$, so that

$$Z(0) = (-1)^k j R_c$$

The input impedance is thus a reactance of magnitude R_c, and is inductive or capacitive according to the sign of $(-1)^k$. In the same way, if $z_l = \infty$, $z(0) = (-1)^{k+1}j$, so that

$$Z(0) = (-1)^{k+1} j R_c$$

is again a reactance of magnitude R_c, and inductive or capacitive depending on k.

These two special cases are put to use in some measurement bridges, such as the General Radio type 1607A, with which the transfer functions of one-port and active two-port networks can be measured over the band 50–1500 MHz.

9.6 *Very short lines*

If the length is much less than a wavelength, $l \ll \lambda$, then $\tan \beta l \approx \beta l$ and

$$z(0) \approx \frac{z_l + j\beta l}{1 + j\beta l z_l} = \frac{z_l + j(2\pi l/\lambda)}{1 + j z_l (2\pi l/\lambda)}$$

For the special case of a short-circuited line,

$$z(0) = j\frac{2\pi l}{\lambda} = j2\pi f(\mathrm{LC})^{1/2} l = j\omega(\mathrm{LC})^{1/2} l$$

which is to say

$$Z(0) = j\omega(\mathrm{LC})^{1/2} l \mathrm{R}_c = j\omega(\mathrm{LC})^{1/2} l(\mathrm{L}/\mathrm{C})^{1/2} = j\omega \mathrm{L} l$$

The line thus appears as an inductance Ll.

For an open-circuited line, $z_l = \infty$, and

$$z(0) = \frac{1}{j(2\pi l/\lambda)} = \frac{1}{j2\pi f(\mathrm{LC})^{1/2} l} = \frac{1}{j\omega(\mathrm{LC})^{1/2} l}$$

which is to say

$$Z(0) = \mathrm{R}_c\, z(0) = \left(\frac{\mathrm{L}}{\mathrm{C}}\right)^{1/2} z(0) = \left(\frac{\mathrm{L}}{\mathrm{C}}\right)^{1/2} \frac{1}{j\omega(\mathrm{LC})^{1/2} l} = \frac{1}{j\mathrm{C}\omega l}$$

This line thus appears as a capacitance Cl.

10. Conclusion

The discussions in this chapter have been brief. They should be considered as simply summaries of material developed in much greater detail elsewhere.

They have been included in this book to demonstrate to the reader that the calculations used in this chapter are identical to those which are used to build the theory of lines in the pulse regime. We hoped thus to convince the reader that lines in the pulse regime can be studied independently of the sinusoidal regime. This is in fact the case for all pulse calculations in electronics, which can be carried out independent of, but alongside, the corresponding calculations for the sinusoidal steady state, both having their common source in the calculations of general electronics.

CHAPTER VI

Solutions to Exercises

Exercises for Chapter II

Exercise 1

Let the positive direction for the current be the same in the two lines (Fig. 157). Because of the short-circuit loads, the reflection coefficient at the load for the two lines is

$$\Gamma_l = \frac{0 - R_c}{0 + R_c} = -1$$

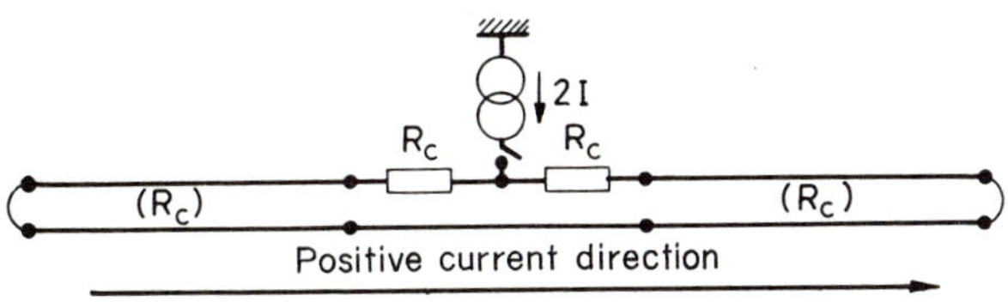

Fig. 157. Circuit for Exercise 1 of Chapter II.

so that

$$v_r = \Gamma_l v_i = -v_i; \qquad i_r = -\Gamma_l i_i = i_i$$

That is, the voltage is reflected with change of sign, and the current is reflected without change of sign. This is diagrammed in Fig. 158.

The input reflection coefficient for the two lines is

$$\Gamma_0 = \frac{3R_c - R_c}{3R_c + R_c} = \frac{1}{2}$$

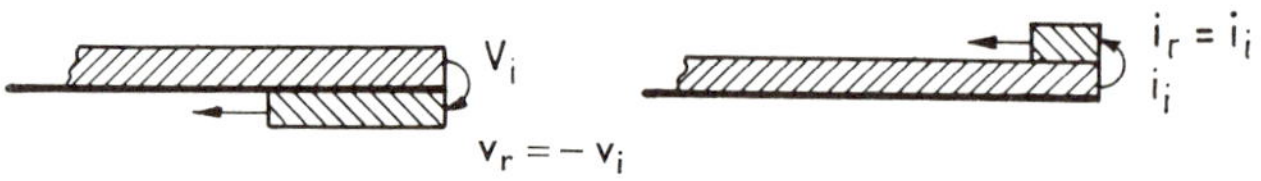

Fig. 158. Voltage and current reflections in circuit of Fig. 157.

since the reflected wave v_r, on returning to the input, sees in series the two lumped resistances R_c and the characteristic resistance R_c of the other line. Thus the wave reflected at the input is

$$v_r = \Gamma_0 v_i = \tfrac{1}{2} v_i; \qquad i_r = -\Gamma_0 i_i = -\tfrac{1}{2} i_i$$

and the transmitted wave is

$$v_t = v_r + v_i = \tfrac{1}{2} v_i + v_i = \tfrac{3}{2} v_i; \qquad i_t = i_i + i_r = i_i - \tfrac{1}{2} i_i = \tfrac{1}{2} i_i$$

These are diagrammed in Fig. 159.

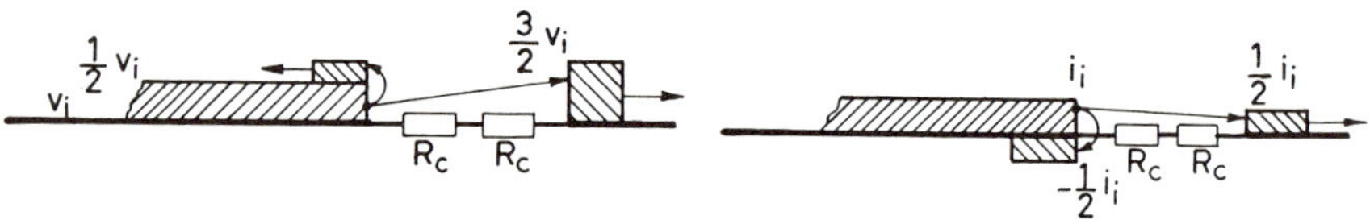

Fig. 159. Wave reflections at the midpoint of the circuit of Fig. 157.

Energy is conserved in this process at the input, since

$$|v_r i_r| + |v_t i_t| = \left|\frac{v_i}{2}\,\frac{i_i}{2}\right| + |\tfrac{3}{2} v_i\, \tfrac{1}{2} i_i| = v_i i_i (\tfrac{1}{4} + \tfrac{3}{4}) = v_i i_i$$

The currents evolve as diagrammed in Fig. 160. Conditions in the time interval $(2\tau, 3\tau)$ are the same as those in the interval $(0, \tau)$, but with opposite signs for all quantities. Thus the current evolution is periodic, with period 4τ. From Fig. 160, it is evident that the currents in the lumped resistances R_c are always identically zero, for $t > 0$. Thus these elements do not dissipate energy, and hence the line never "runs down" to the zero steady state, but continues to carry oscillating currents for all time.

Note that the above conclusions hold also if the lumped resistances R_c are replaced by resistances $\varrho \neq R_c$. In that case, we would have

$$\Gamma_l = \frac{(2\varrho + R_c) - R_c}{(2\varrho + R_c) + R_c} = \frac{K - 1}{K + 1}$$

where K is defined by

$$2\varrho + R_c = K R_c$$

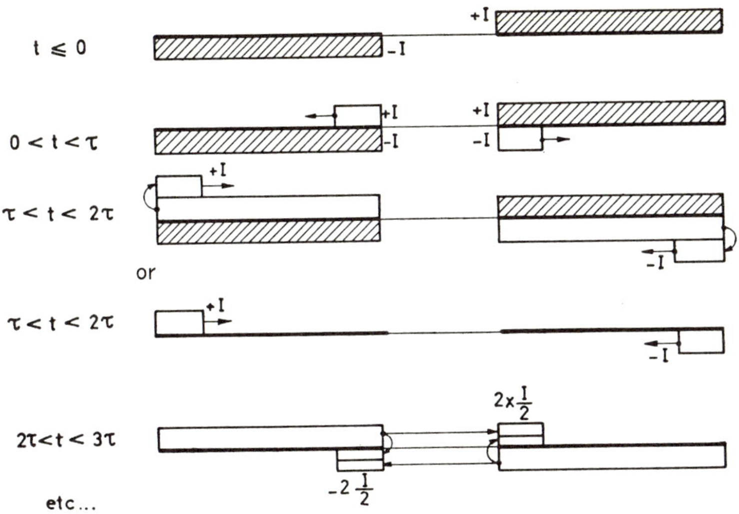

Fig. 160. Successive current waves on the line of Fig. 157.

But this would change nothing essential in the argument, since the oscillations would be of the same character, and the lumped resistances would still carry identically zero currents.

There is a flaw in the above problem, even if one is willing to grant the existence of a lossless line. We have supposed that at $t = 0$ there was established a steady state with a current I in each line. But this is not possible, because the current evolution for establishing current in the line is of the same nature as that we have investigated in the case that the source is removed. With lossless lines, the currents would thus oscillate indefinitely, and the assumed steady state could not have been established. See Fig. 161. In

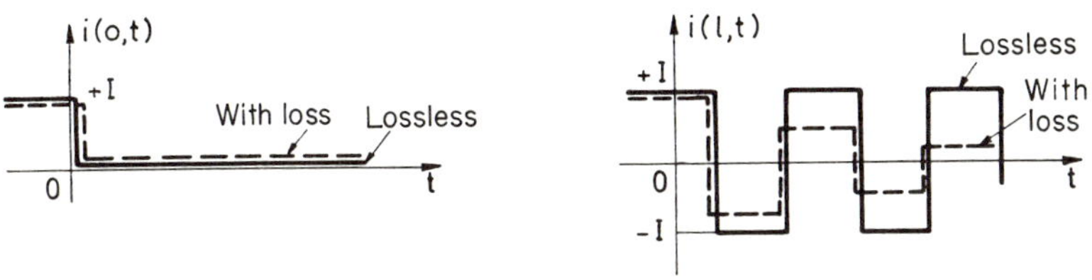

Fig. 161. Source and load currents in line of Fig. 157.

practice, however, the losses in the line allow the equilibrium state to be attained for $t < 0$. The case of the lossless line treated above is only an approximation to reality. The approximation can easily be corrected by introducing the attenuation due to losses into the argument. This yields for each line, after the source is removed, the diagrams of Fig. 161.

Exercise 2

We have the circuit of Fig. 162 to analyze. The line voltage evolves as in Fig. 163. At the instant $2\tau - \varepsilon$, the voltage at A reaches the value 2E, and the diode no longer conducts. We must now determine the wave (v_i', i_i')

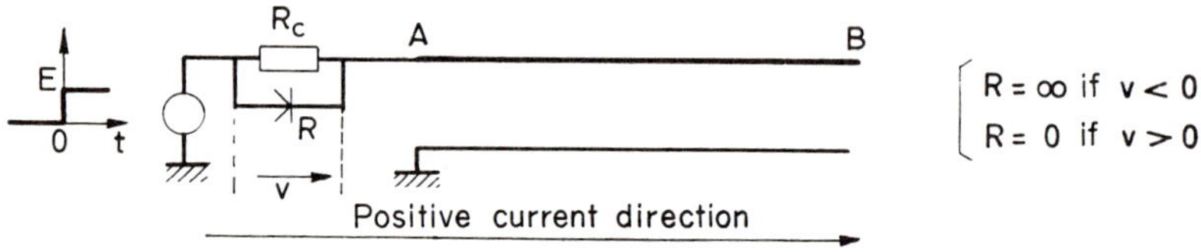

Fig. 162. Circuit for Exercise 2 of Chapter II.

reflected at the input at time 2τ. Even though the source is matched, having in effect internal resistance R_c, the reflected wave is not zero, since the voltage at point A is not the voltage of the generator.

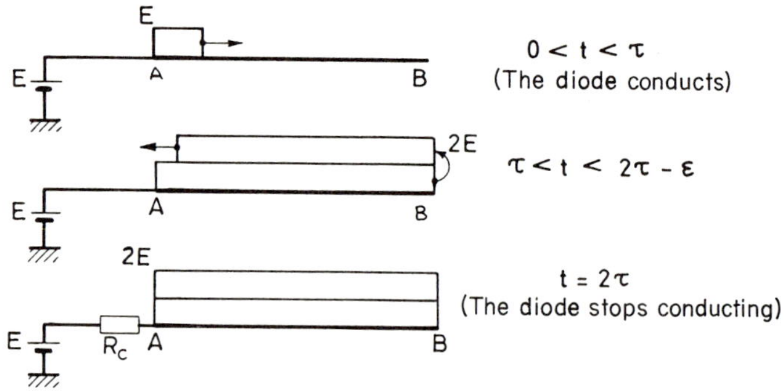

Fig. 163. Voltage waves in circuit of Fig. 162.

We have

$$v_A = v_i + v_r + v_i'; \qquad i_A = i_i + i_r + i_i'$$

But

$$v_A = E - R_c i_A$$

(according to the positive direction chosen for the current), and

$$v_i = E; \qquad v_r = E; \qquad i_i = \frac{v_i}{R_c}; \qquad i_r = -\frac{v_r}{R_c}$$

Thus

$$2E + v_i' = E - R_c i_i' = E - v_i'$$

from which $v_i' = -E/2$, so that

$$v_A = v_i' + v_i + v_r = 3E/2$$

The further evolution of the voltage is thus as in Fig. 164.

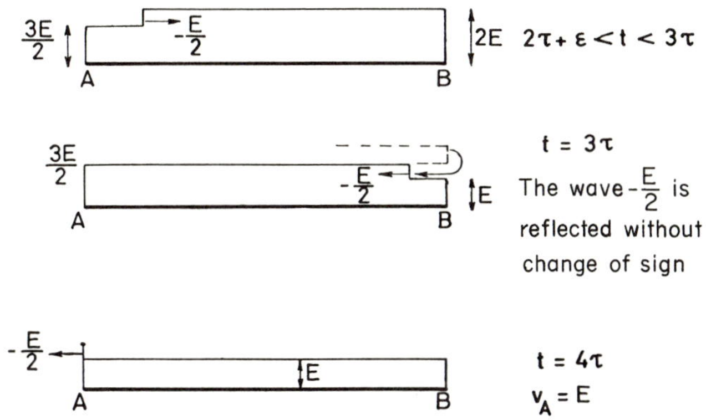

Fig. 164. Voltage waves in circuit of Fig. 162 after the diode ceases to conduct.

At times $t \geq 4\tau$, v_A is equal to E, since

$$v_A = v_i + v_r + v_i' + v_r' + v_i''$$
$$v_A = E - R_c i_A = E - R_c(i_i + i_r + i_i' + i_r' + i_i'')$$

which yields, upon replacing successive waves by their values found above,

$$E + v_i'' = E - R_c i_i'' = E - v_i''$$

or finally

$$v_i'' = 0$$

We also have

$$v_B = v_i + v_r + v_i' + v_r' = E$$

Thus the complete evolutions of the input and output voltages are as in Fig. 165.

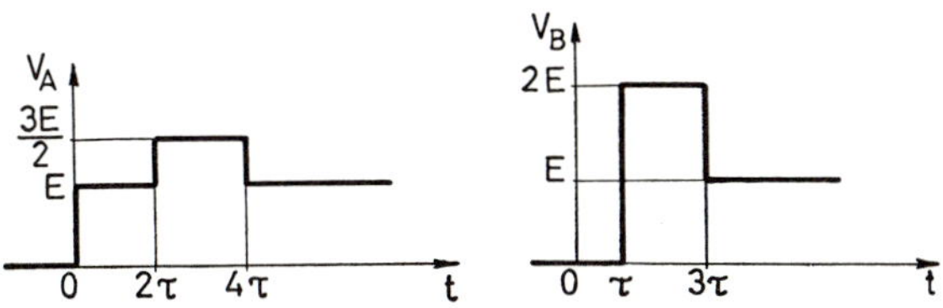

Fig. 165. Input and output voltages for the circuit of Fig. 162.

Exercise 3

Since the circuit of Fig. 166 and all its components are linear, the principle of superposition can be applied when convenient. We will take as time origin $t = 0$ the instant at which a voltage first appears at point A. To calculate that voltage, consider that the incident wave arriving at A sees

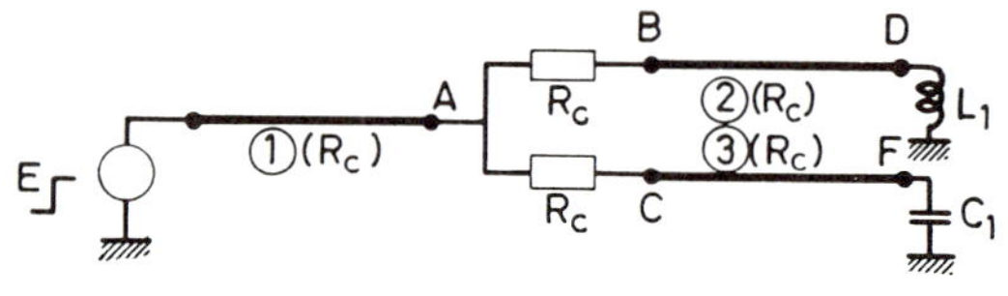

Fig. 166. Circuit for Exercise 3 of Chapter II.

two resistances $2R_c$ in parallel, equivalent to a resistance R_c. Thus line 1 is matched. Then according to Section 2.3, line 1 is equivalent to a generator with emf $2v_i = 2E$ and internal resistance R_c. Hence $v_A = E$ initially (Fig. 167).

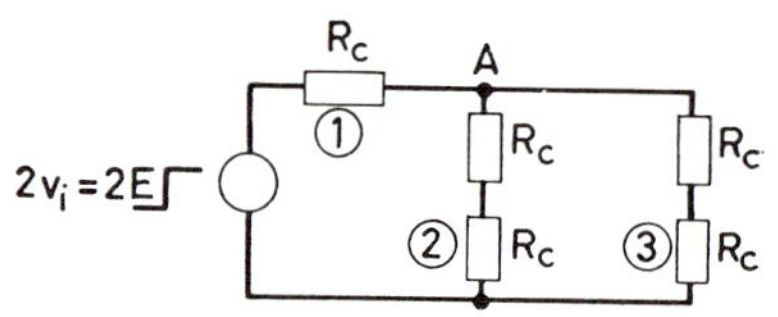

Fig. 167. Initial equivalent circuit for line of Fig. 166.

Let us now consider separately the inductive and capacitive branches (Fig. 168). Both these problems have already been treated in Chapter II. It is necessary to require that the time constants L_1/R_c and R_cC_1 be the same. Then superposing the diagrams of Fig. 168, it is seen that v_B and v_C

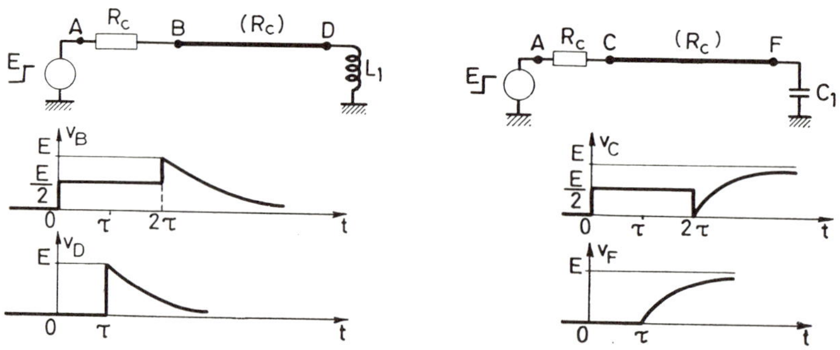

Fig. 168. Voltage evolutions for the two branches of the circuit of Fig. 166.

compensate exactly, and maintain the voltage at A constant and equal to E after as well as before $t = 2\tau$. Thus everything evolves as if the source were feeding a matched line.

Exercise 4

We have the circuit of Fig. 169. For $0 < t < 2\tau$, at A we can write

$$v_A = RI = R_c i \qquad \text{with} \quad I_1 = I + i$$

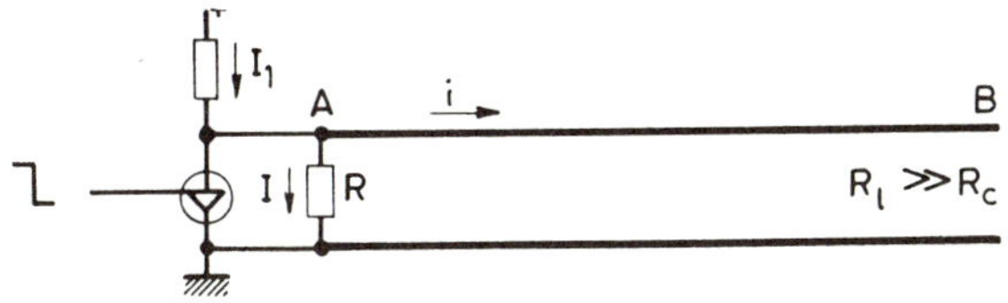

Fig. 169. Circuit for Exercise 4 of Chapter II.

from which

$$v_A = \frac{RR_c}{R + R_c} I_1 = v_0$$

Here v_0 is the amplitude of the incident wave v_i which is launched down the line. The corresponding current is

$$i_i = \frac{v_i}{R_c} = \frac{RI_1}{R + R_c}$$

At B, for $\tau < t < 3\tau$, there is voltage reflection without change of sign, $v_r = v_i = v_0$, and current reflection with change of sign, $i_r = -i_i$. Thus

$$v_B = 2v_0; \qquad i_B = 0$$

At A, for $2\tau < t < 4\tau$, the reflected wave is itself reflected, to produce new incident waves v_i', i_i', and

$$v_A = v_i + v_r + v_i' = RI = R(I_1 - i_i - i_r - i_i') = R(I_1 - i_i')$$

$$R_c(i_i - i_r + i_i') = R(I_1 - i_i')$$

From these

$$i_i' = \frac{R(R - R_c)}{(R + R_c)^2} I_1 = -\Gamma_0 i_r; \qquad v_i' = R_c i_i' = \Gamma_0 v_r$$

where

$$\Gamma_0 = \frac{R - R_c}{R + R_c}$$

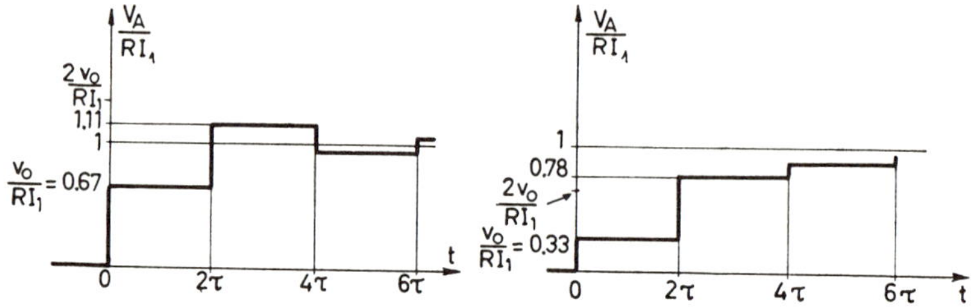

Fig. 170. Input voltage for line of Fig. 169, for two cases of source resistance.

If $R > R_c$, i_i', v_i', and Γ_0 are positive, and v_A will be greater than $2v_0$:

$$v_A = v_0(2 + \Gamma_0); \qquad i_A = \frac{v_0}{R_c}\Gamma_0$$

If $R < R_c$, i_i', v_i', and Γ_0 are negative, and v_A will be less than $2v_0$.

At B, for $3\tau < t < 5\tau$, the voltage will be

$$v_B = v_i + v_r + v_i' + v_r' = 2v_i + 2v_i' = 2v_0(1 + \Gamma_0)$$

At A, for $4\tau < t < 6\tau$, we will have

$$v_A = v_0(2 + 2\Gamma_0 + \Gamma_0^2); \qquad i_A = \frac{v_0}{R_c}\Gamma_0^2$$

and so forth. The final value of both v_A and v_B is

$$RI_1 = \frac{2v_0}{1 - \Gamma_0}$$

The waveforms of the voltage and current are similar to those found in the example in Section 3.2. For example, for the voltage v_A we obtain the two graphs of Fig. 170.

Exercise 5

We have the circuit of Fig. 171. The generator voltage is

$$U = \frac{50 + 10}{50} = 1.2 \text{ volts}$$

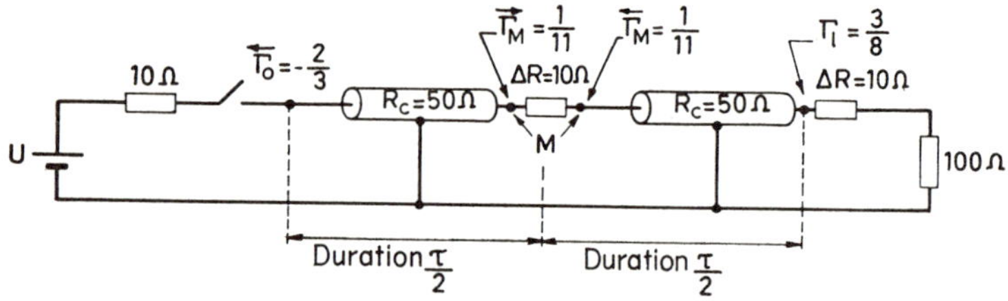

Fig. 171. Circuit for Exercise 5 of Chapter II.

The various reflection coefficients are

$$\overleftarrow{\Gamma}_0 = \frac{10 - 50}{10 + 50} = -\frac{2}{3} = -0.667$$

$$\overrightarrow{\Gamma}_{\mathrm{M}} = \frac{(50 + 10) - 50}{(50 + 10) + 50} = +\frac{1}{11} = +0.091$$

$$\Gamma_l = \frac{(10 + 100) - 50}{(10 + 100) + 50} = +\frac{3}{8} = +0.375$$

We assume that the element $\Delta\mathrm{R}$ is of negligible length.

In constructing the table of Fig. 172, account should be taken of the attenuation introduced by the two resistances $\Delta\mathrm{R}$. In effect, the step leaving the first section of line sees an impedance $\Delta\mathrm{R} + \mathrm{R}_c = 10 + 50 = 60\,\Omega$, and only a fraction $50/60 = 5/6$ is transmitted. There is thus an attenuation coefficient $\mathrm{R}_c/(\Delta\mathrm{R} + \mathrm{R}_c)$ (the same in the two directions), so that

$$\overrightarrow{r}_1 = \overleftarrow{r}_1 = \frac{50}{50 + 10} = 0.833$$

In the same way, at the output of the line, the attenuation coefficient is

$$\overrightarrow{r}_2 = \frac{100}{100 + 10} = 0.909$$

The final state of the line has for voltages at the input, in the middle, and at the output

$$\mathrm{V}_{0\infty} = 1.2 \times \frac{120}{130} = 1.11 \text{ volt}$$

$$\mathrm{V}_{\mathrm{M}\infty} = 1.2 \times \frac{110}{130} = 1.015 \text{ volt}$$

$$\mathrm{V}_{1\infty} = 1.2 \times \frac{100}{130} = 0.923 \text{ volt}$$

It is now possible to fill the table of Fig. 172. For example, the unit step at the input is transmitted to the middle of the line, and there splits into a reflected wave of amplitude 0.091 and a transmitted wave of amplitude $1 + 0.091 = 1.091$, of which however only the fraction r_1 is actually transmitted, i.e., the transmitted wave has amplitude $1.091 \times r_1 = 1.091 \times 0.833 = 0.909$. This value is circled in the table, since it is the voltage at the output of the first line section up till time $3\tau/2$.

The steps of amplitudes 0.091 and 0.909 are then transmitted in the usual way. When these two waves again return to the middle, with amplitudes

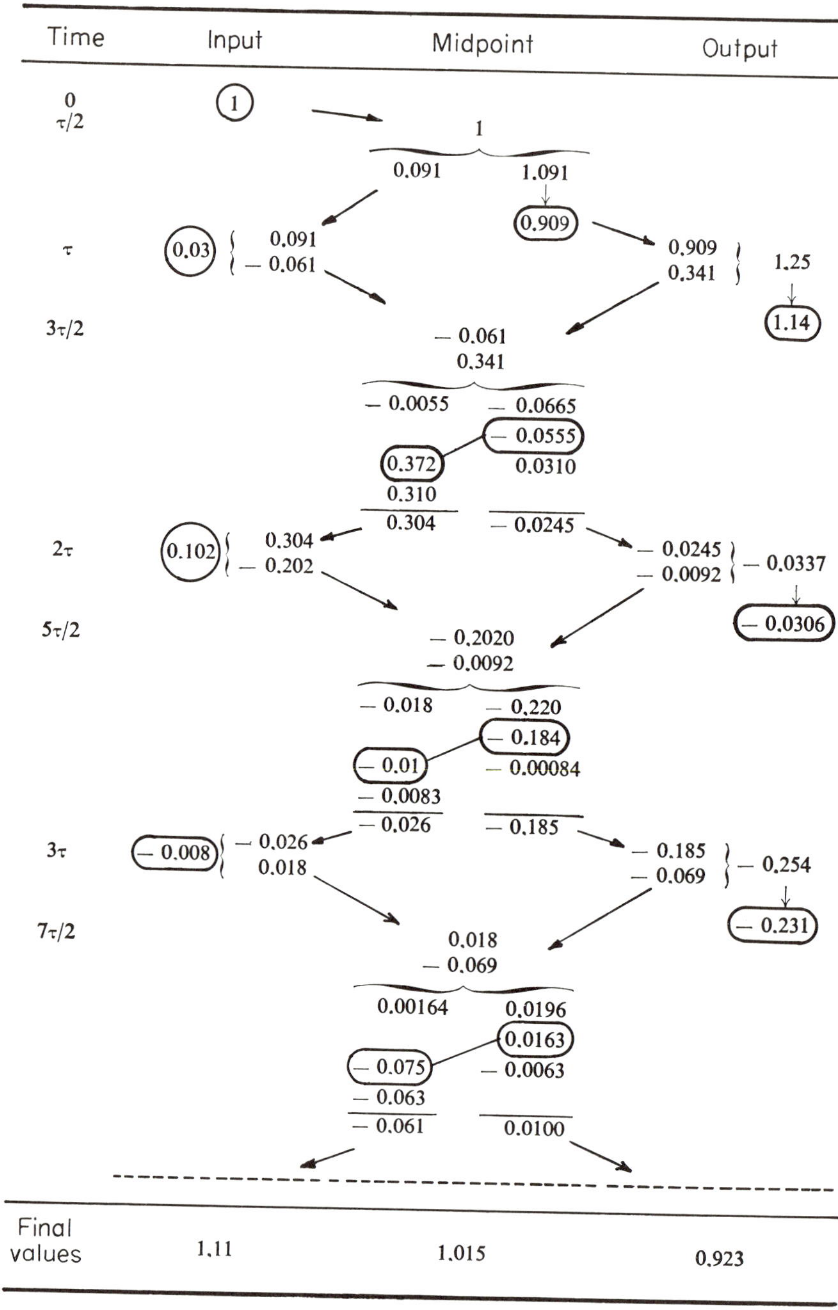

Fig. 172. Table of wave amplitudes for circuit of Fig. 171.

-0.061 and 0.341, the transmitted waves are again reduced, i.e., -0.0665 and 0.372 are reduced by the same factor 0.833. The values -0.0555 and 0.372 are circled, since they are the wave amplitudes at the midpoint. Note that the output wave amplitude is not 1.25, but rather $1.25 \times 0.909 = 1.14$, on account of the reduction coefficient r_2.

The table is completed in the same way (Fig. 172).

Exercise 6

The problem is to analyze the structure in Fig. 173. In this case

$$U = 1.2 \text{ volts}$$

$$\overleftarrow{\Gamma}_0 = -\tfrac{2}{3}; \qquad \overrightarrow{\Gamma}_D = \overleftarrow{\Gamma}_D = -\tfrac{1}{3}; \qquad \overrightarrow{\Gamma}_{l_1} = -1; \qquad \overrightarrow{\Gamma}_{l_2} = 1$$

Since the propagation times are not the same for all lines, it is necessary

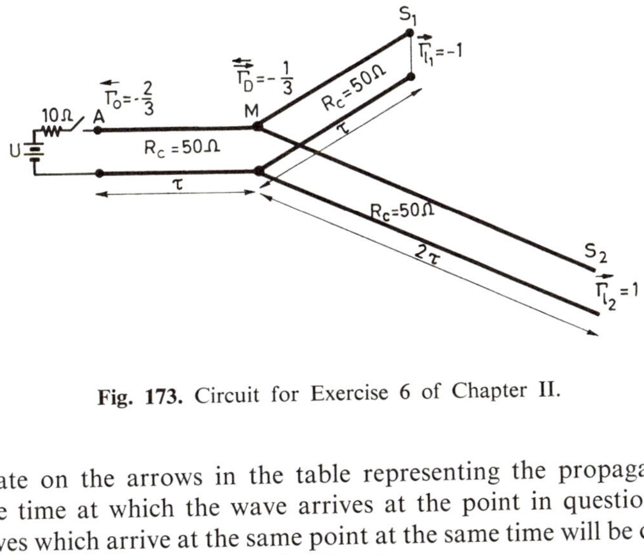

Fig. 173. Circuit for Exercise 6 of Chapter II.

to indicate on the arrows in the table representing the propagation of a wave the time at which the wave arrives at the point in question, so that only waves which arrive at the same point at the same time will be combined.

The completed table is shown in Fig. 174.

Exercises for Chapter III

Bergeron's method can be summarized as follows. First, a wave carrying an observer is launched from the point corresponding to the initial state (v_0, i_0). At the first point of discontinuity (an open-circuited or short-circuited line end, or a change in characteristic impedance), the relation

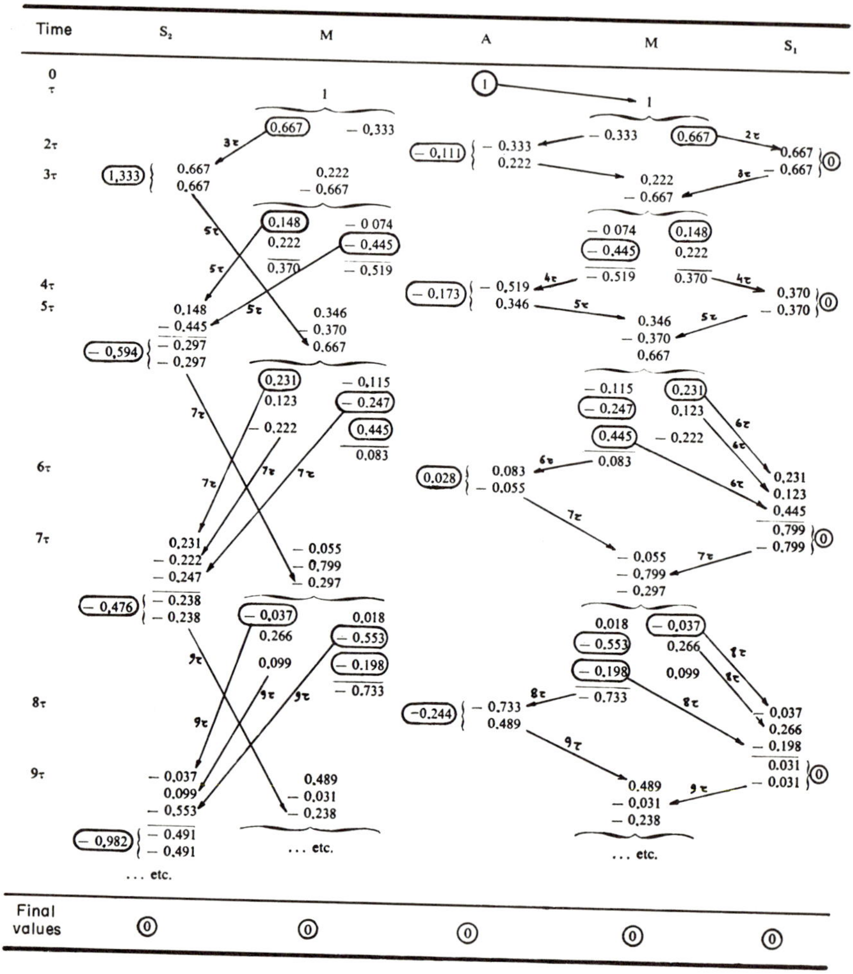

Fig. 174. Table of wave amplitudes for circuit of Fig. 173.

between v and i is known. One or two new waves are emitted, along with observers. The travels of the observers are followed. Each observer is related to new observers whenever there is refraction or reflection at a discontinuity.

Exercise 1

We wish to consider the discharge of a line when it is suddenly short-circuited at $t = 0$, as in Fig. 175, the line having been previously charged to a voltage E. At $t = 0$, since point A is grounded, the voltage v_A suddenly drops to zero. This corresponds to launching an incident wave $-E$, with

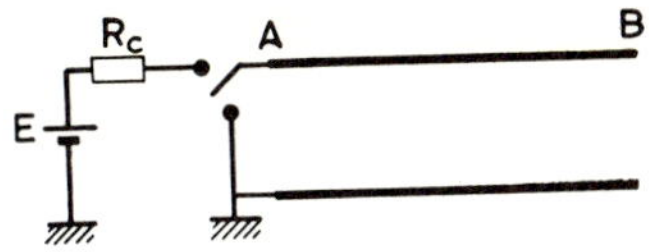

Fig. 175. Circuit for Exercise 1 of Chapter III.

corresponding current wave $-E/R_c$. This initial state at A is represented on the diagram in Fig. 176 by the point $(-E/R_c, 0)$:A(0). The incident wave is represented by the line with slope $-R_c$ passing through the point A(0).

Point B being an open circuit, the current is necessarily zero. Thus the state at point B, at time $t = \tau$, the propagation time, is located at the

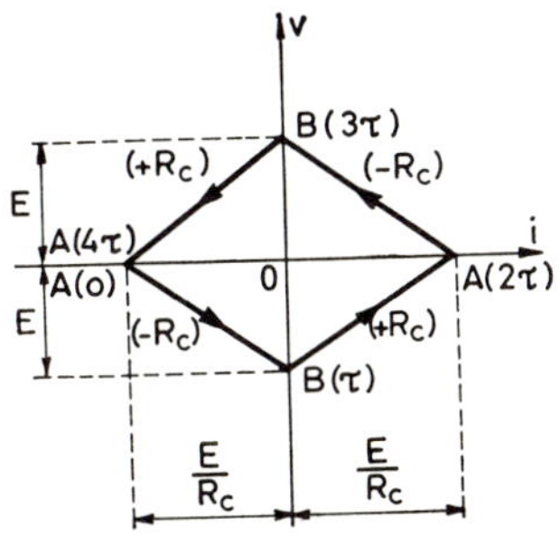

Fig. 176. State diagram for circuit of Fig. 175.

intersection of the line just drawn, and the v axis. This point is marked $B(\tau)$ on the diagram.

At B a reflected wave appears, represented by the line of slope R_c passing through the point $B(\tau)$. This wave arrives at A at time 2τ, at which point the voltage is zero, because of the short circuit. Thus the state of A at time 2τ is represented by the intersection of the line just drawn, with the i axis. The evolution then proceeds in the same way. The state diagram

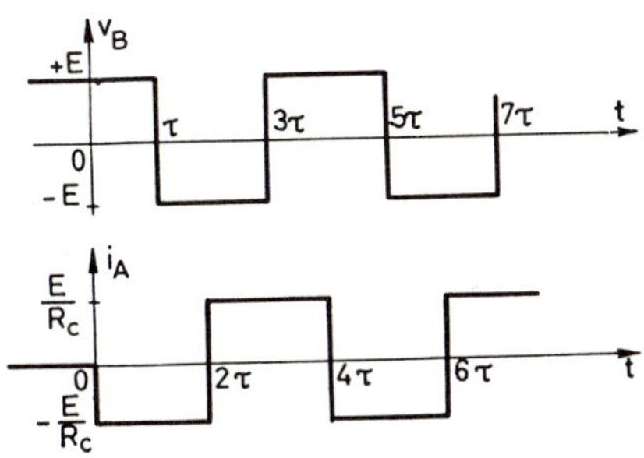

Fig. 177. Output voltage and input current for circuit of Fig. 175.

is thus a lozenge, and is traversed indefinitely with period 4τ. The curves of $i_A(t)$ and $v_B(t)$ shown in Fig. 177 follow at once. The oscillations persist indefinitely because the line is lossless. If the line had ohmic losses, the currents and voltages would gradually die out to zero.

Exercise 2

The solution is that shown in Fig. 178.

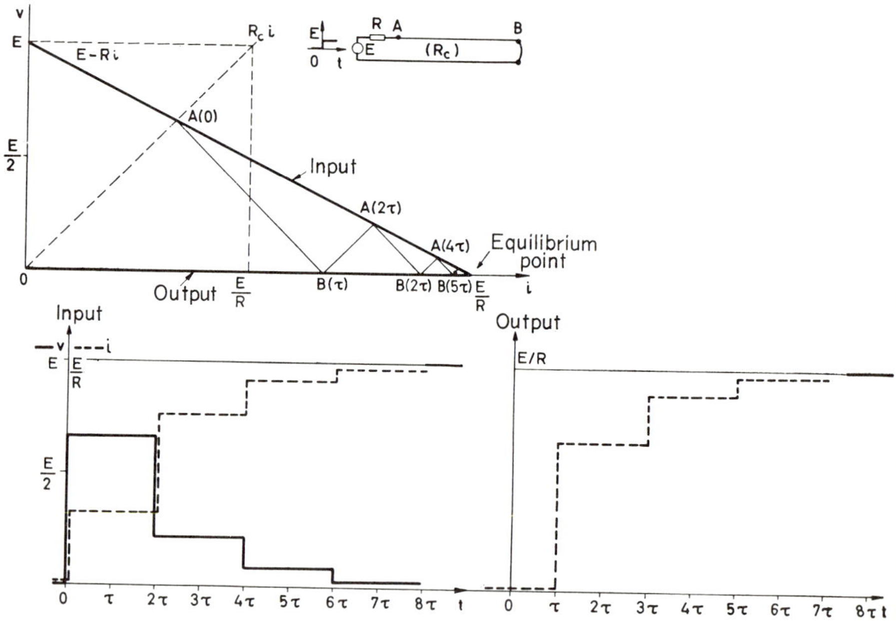

Fig. 178. Circuit, Bergeron diagram, and current and voltage waves for Exercise 2 of Chapter III.

Exercise 3

The circuit is that of Fig. 179, the voltage E being applied at $t = 0$, and the diode being ideal. At $t = 0$, an observer leaves point A, corresponding to a voltage wave E and current wave E/R_c, since R_c is the input impedance of the line, and the diode is conducting (Fig. 180).

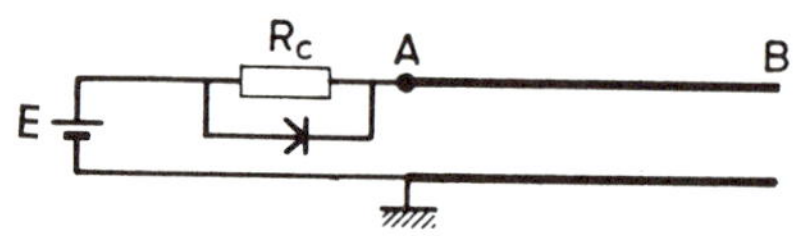

Fig. 179. Circuit for Exercise 3 of Chapter III.

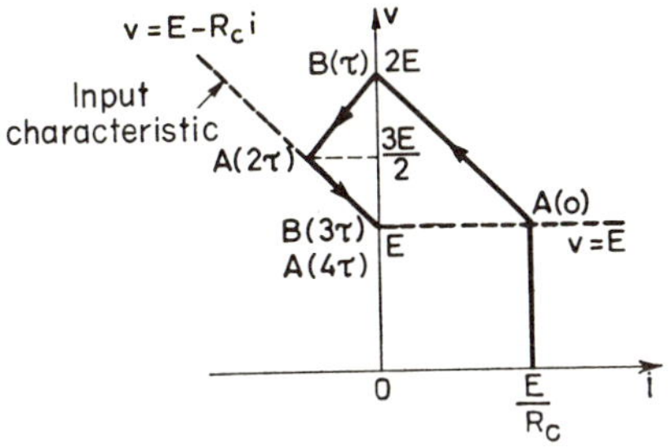

Fig. 180. State diagram for circuit of Fig. 179.

As the wave propagates from point A of the line to B, the line with slope $-R_c$ passing through A(0) is traced. At B the current is zero, because of the open circuit. Thus the point B(τ) on the diagram has coordinates $v = 2E$, $i = 0$.

A reflected wave is created at B, and departs for A. The line with slope R_c passing through B(τ) is traced. When this wave arrives at A, it encounters either a resistance R_c and the emf E, if v_A is greater than E, or else the voltage E alone. Thus the point A(2τ) is at the intersection of the line just drawn and one or the other of the input circuit characteristic lines $v = v_A = E - R_c i$, $v = E$, depending on v_A. This combined input characteristic is shown as the dashed line in Fig. 180. The point of intersection A(2τ) is at $v = 3E/2$, $i = E/2R_c$.

This wave is reflected at A and departs for B, tracing the line with slope $-R_c$ passing through A(2τ). At B, the current must again be zero, so that the point B(3τ) is on the v axis.

A reflected wave now departs for A along the line with slope R_c through B(3τ). Upon arriving at A, this wave again encounters the input circuit, and its representative point must lie on the dashed curve of Fig. 180. The only possibility is that the representative point has not moved at all in the (v, i) plane, and that the points A(4τ) and B(3τ) are identical. Further, the reflected wave launched towards B must be represented by a point moving along the line with slope $-R_c$ passing through A(4τ), and since the current at B is still necessarily zero, the state B(5τ) is the intersection of this line with the v axis. But again the only possibility is that the re-

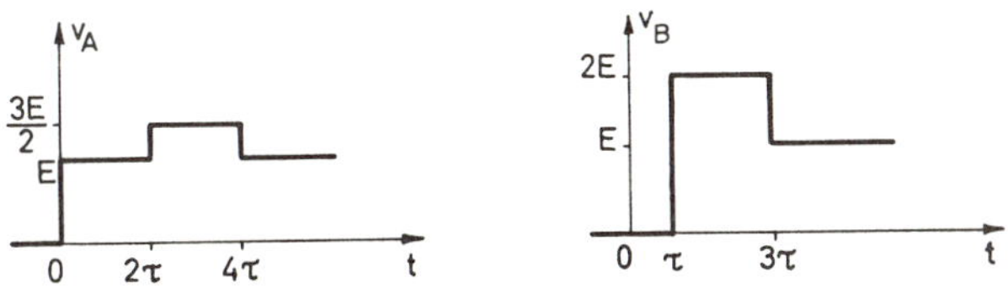

Fig. 181. Input and output voltages for line of Fig. 179.

presentative point does not move at all, and B(5τ) is identical with A(4τ). The diagram ceases to evolve, the points B(3τ), A(4τ), B(5τ), A(6τ), ... being all identical. The voltage versus time graphs of Fig. 181 result.

Exercise 4

No solution is provided.

Exercise 5

The solution is indicated in Fig. 182.

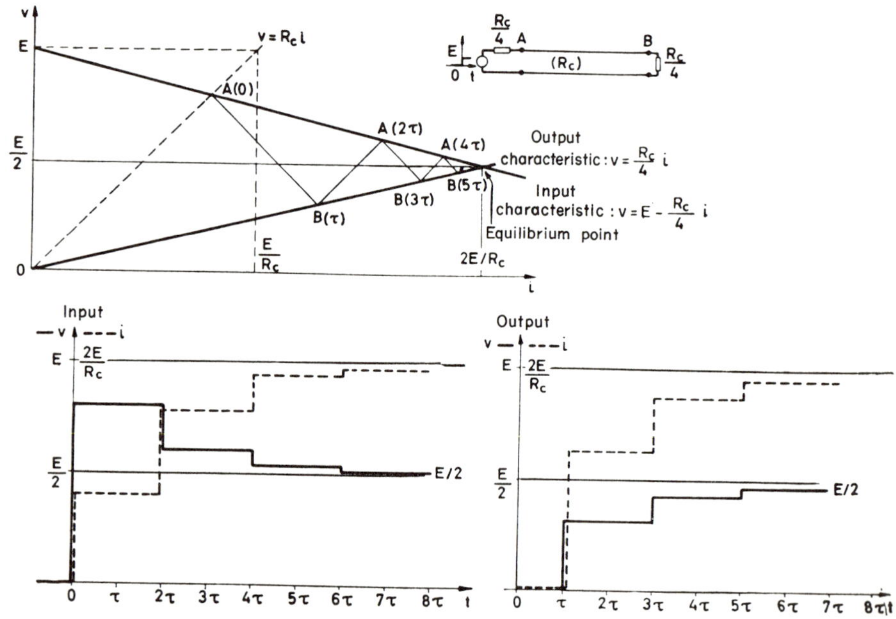

Fig. 182. Circuit, Bergeron diagram, and current and voltage waveforms for Exercise 5 of Chapter III.

Exercise 6

At the initial time $t = 0$, v_A jumps from $-E$ to 0, and a voltage wave E and current wave E/R_c are launched down the line. This gives the initial point A(0) in the diagrams of Fig. 183. The input and output states B(τ), A(2τ), ... are the intersections of the appropriate characteristic lines of slopes $+R_c$ and $-R_c$ with the characteristic curves of the input and output circuits. These latter are, respectively, the i axis, since $v = 0$ at the input, and the volt-ampere characteristic of the tunnel diode, taking account of the bias $-E$.

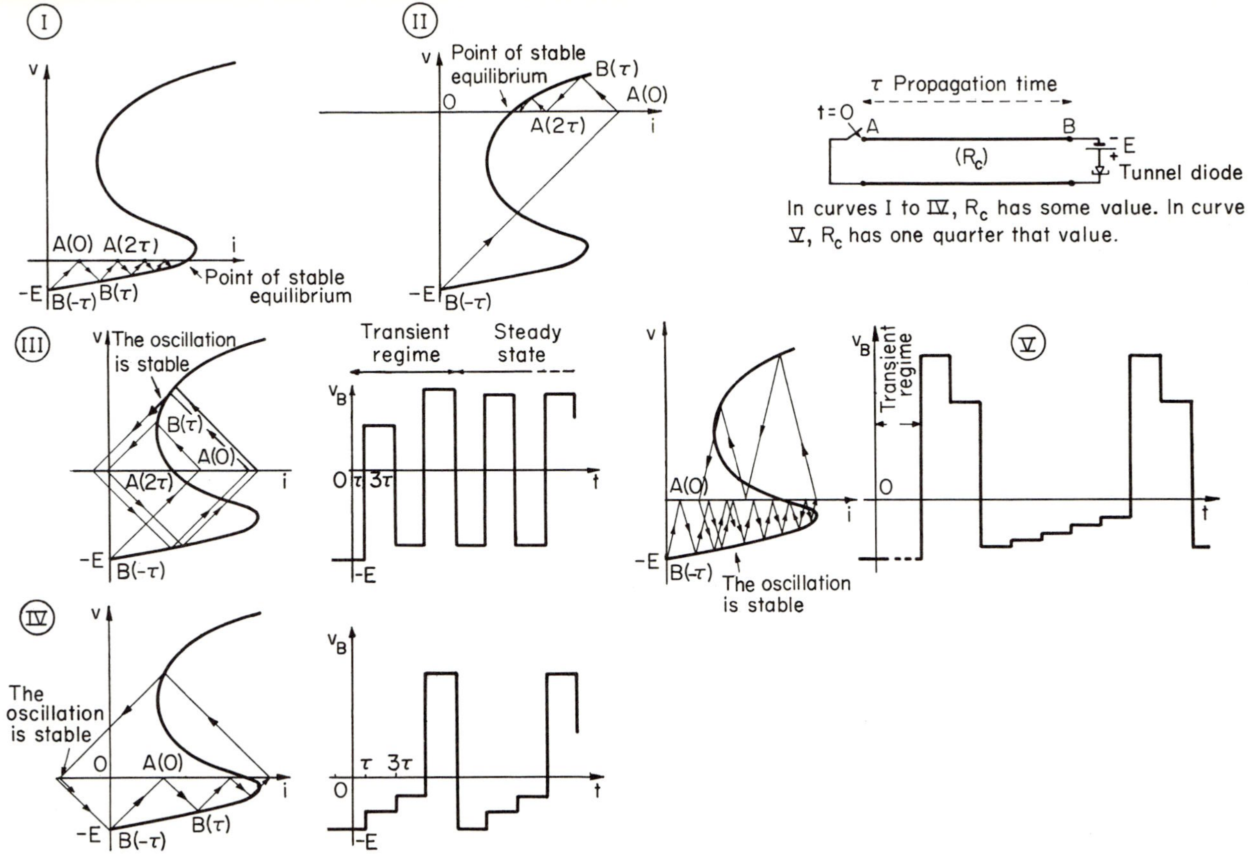

Fig. 183. Circuit, state diagrams, and output voltage waveforms for Exercise 6 of Chapter III.

Graphs 1 and 2 of Fig. 183 show two examples in which a point of stable equilibrium is attained after a short initial transient time. Graphs 3–5 show cases in which a steady oscillation is established after a transient time. In case 5, the characteristic resistance of the line is only a fourth that used for cases 3 and 4.

To our knowledge, operation of a tunnel diode in this manner was accomplished for the first time in March 1961 by Chapouille, of the Compagnie des Machines Bull. Independent work has been done by Nagumo and Shimura of the University of Tokyo.† These latter workers describe a method very near to that of Bergeron, a fact which led to a correspondence item in the *Proceedings*.‡

Exercise 7

The current and voltage at point M of the circuit of Fig. 184 are to be found. Point M is $\frac{1}{3}$ the distance down a line of length l, with characteristic resistance R_c, fed by a voltage step and loaded by $R_l = R_c/2$.

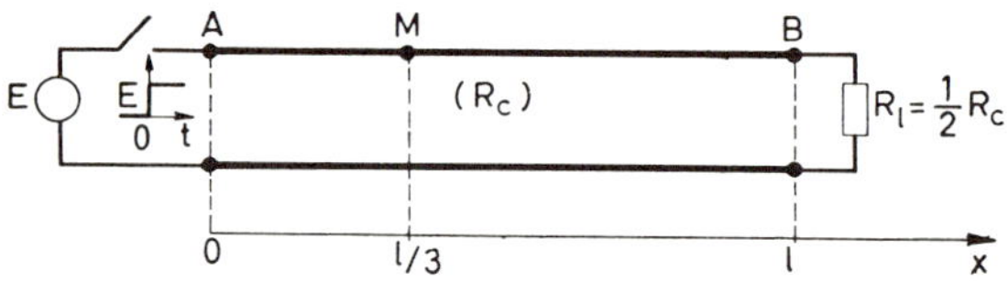

Fig. 184. Circuit for Exercise 7 of Chapter III.

The states at A and B, the input and output, were found in Section 2.1,1 of Chapter III (the case $R_l < R_c$). The graph shown in Fig. 185 was the result.

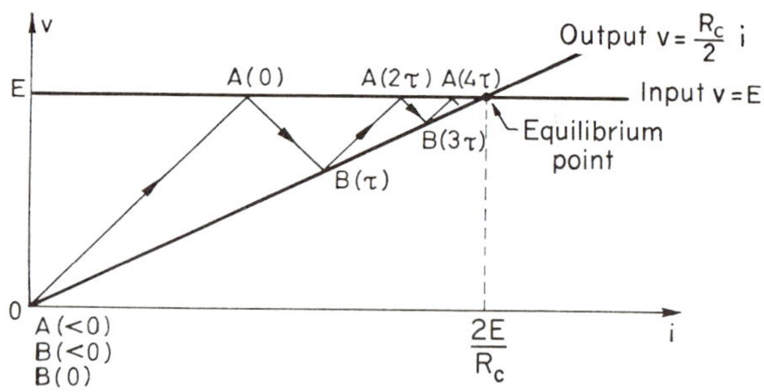

Fig. 185. Input and output states for circuit of Fig. 184.

† See J. Nagumo and M. Shimura, *Proc. IRE* August (1961).

‡ See P. Chapouille and J.-P. Vabre, *Proc. IRE* 2373–2374 (Nov) (1962).

To determine the state at M, the point with $x = l/3$, at time t, an observer O_1 is caused to depart from A at time $t - \tau/3$, and an observer O_2 from B at $t - 2\tau/3$. (Here τ is the propagation time of the line.)

The current, or the voltage, or both, change value at A at times 0, 2τ, ..., $2k\tau$, ..., and at B at times τ, 3τ, ..., $(2k + 1)\tau$, ..., where k is any integer.

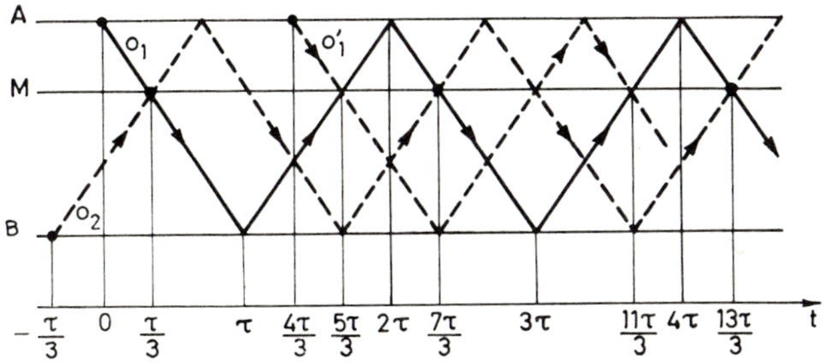

Fig. 186. Timetable for observers in circuit of Fig. 184.

Thus changes at point M occur at times $\tau/3$, $2\tau + \tau/3$, $4\tau + \tau/3$, ... due to changes at A, and at times $\tau + 2\tau/3$, $3\tau + 2\tau/3$, $5\tau + 2\tau/3$, ... due to changes at B. These times are indicated on the "railroad" diagram in Fig. 186. The table below can now be set up, indicating the times of interest at points A and B, and the corresponding times at which changes occur at M.

Times $\left(t - \frac{\tau}{3}\right)$ at A	Times $\left(t - \frac{2\tau}{3}\right)$ at B	Times t at M
0		$\frac{\tau}{3}$
	τ	$\frac{5\tau}{3} = \tau + \frac{2\tau}{3}$
2τ		$\frac{7\tau}{3} = 2\tau + \frac{\tau}{3}$
	3τ	$\frac{11\tau}{3} = 3\tau + \frac{2\tau}{3}$
4τ		$\frac{13\tau}{3} = 4\tau + \frac{\tau}{3}$
	5τ	$\frac{17\tau}{3} = 5\tau + \frac{2\tau}{3}$
etc.	etc.	etc.

In the (v, i) diagram of Fig. 187, up till time $\tau/3$, the point representing M remains at the origin, since, as can be seen from Fig. 185, no perturbation reaches M until that time. At the instant $\tau/3$, the state of M changes. The graph of Fig. 186 shows that at this time observer O_1, traveling from A(0), and observer O_2, traveling from $B(-\tau/3)$, reach M. Since $B(-\tau/3)$ is at the origin, the point $M(\tau/3)$ has the same coordinates as A(0). This is shown in Fig. 187.

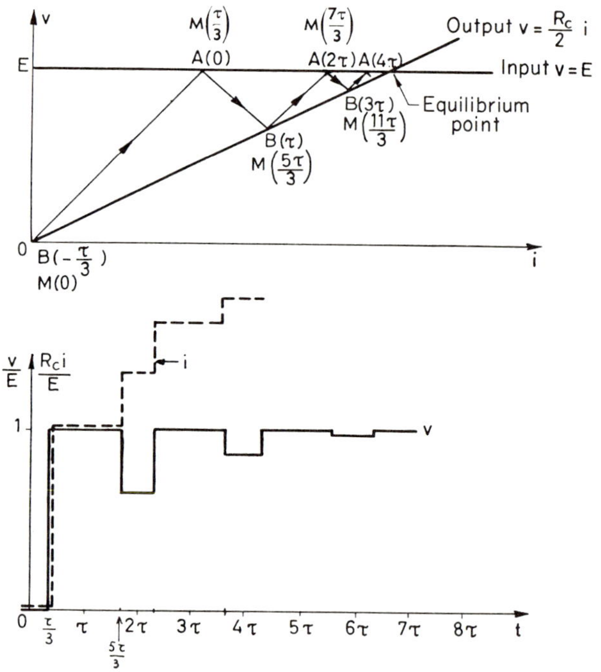

Fig. 187. States, voltage, and current, at point M in circuit of Fig. 184.

The point $M(5\tau/3)$ is at the intersection of the line traversed by an observer O_1' leaving $A(4\tau/3)$, which is the same as A(0), and the line traversed by O_2', who leaves $B(\tau)$. Thus $M(5\tau/3)$ is the same point as $B(\tau)$.

In general, as indicated in Fig. 187, the Bergeron diagram for this problem, the state of point M at any time indicated in the table, is the same as the state of the point A or B at the time indicated in the same line of the table. Thus $M(\tau/3)$ and A(0) are the same, as are $M(5\tau/3)$ and $B(\tau)$, $M(7\tau/3)$ and $A(2\tau)$, etc. This is supported by the general relations of Section 2.2,1 of Chapter III.

Exercises for Chapter IV

Exercise 1

The circuit of interest, and the solution, are shown in Fig. 188. For this case, $\Gamma_0 = -1$, $\Gamma_l = 0$. Formula (IV.5) yields

$$\mathrm{V}(x, p) = \frac{\mathrm{E}}{p} e^{-p\delta x}$$

from which

$$v(x, t) = \mathrm{E}\Upsilon(t - \delta x)$$

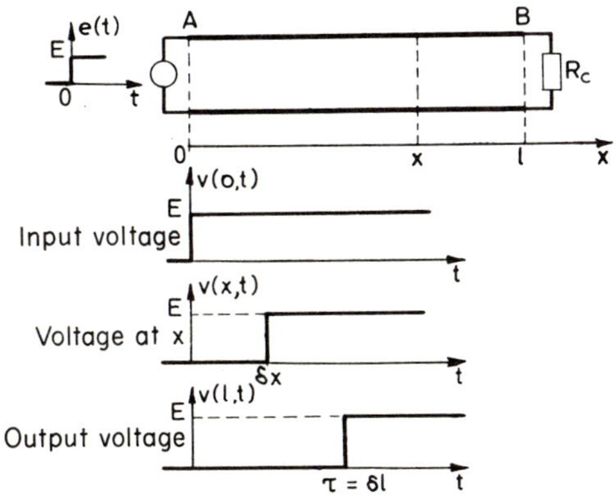

Fig. 188. Circuit and voltage waveforms for Exercise 1 of Chapter IV.

Exercise 2

For the circuit of Fig. 189, we have $\mathrm{Z}_0 = \mathrm{R}_c$, hence $\Gamma_0 = 0$, $\mathrm{Z}_l = \infty$, hence $\Gamma_l = 1$, and $\mathrm{E}(p) = \mathrm{E}/p$. The basic equations then become

$$\mathrm{V}(x, p) = \frac{\mathrm{E}}{2p}\left(e^{-p\delta x} + e^{-p\delta(2l-x)}\right)$$

$$\mathrm{I}(x, p) = \frac{\mathrm{E}}{2p\mathrm{R}_c}\left(e^{-p\delta x} - e^{-p\delta(2l-x)}\right)$$

from which

$$v(x, t) = \frac{\mathrm{E}}{2}\left[\Upsilon(t - \delta x) + \Upsilon(t - \delta(2l - x))\right]$$

$$i(x, t) = \frac{\mathrm{E}}{2\mathrm{R}_c}\left[\Upsilon(t - \delta x) - \Upsilon(t - \delta(2l - x))\right]$$

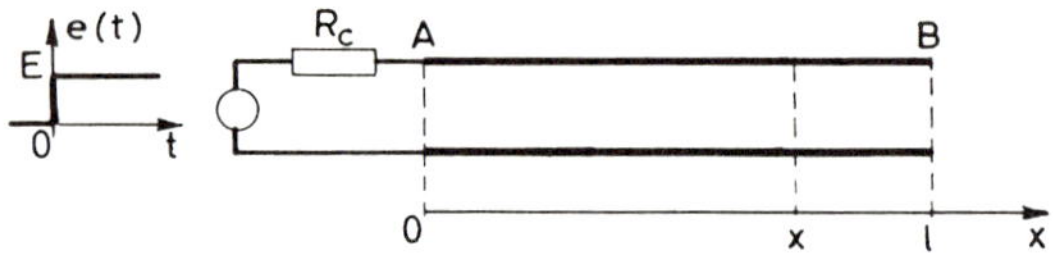

Fig. 189. Circuit for Exercise 2 of Chapter IV.

At the input, $x = 0$, and these equations become

$$v(0, t) = \frac{E}{2} [\Upsilon(t) + \Upsilon(t - 2\delta l)]$$

$$i(0, t) = \frac{E}{2R_c} [\Upsilon(t) - \Upsilon(t - 2\delta l)]$$

For the output, since $x = l$, we have

$$v(l, t) = \frac{E}{2} [\Upsilon(t - \delta l) + \Upsilon(t - \delta l)] = E\Upsilon(t - \delta l)$$

$$i(l, t) = \frac{E}{2R_c} [\Upsilon(t - \delta l) - \Upsilon(t - \delta l)] = 0$$

These results are shown in Fig. 190.

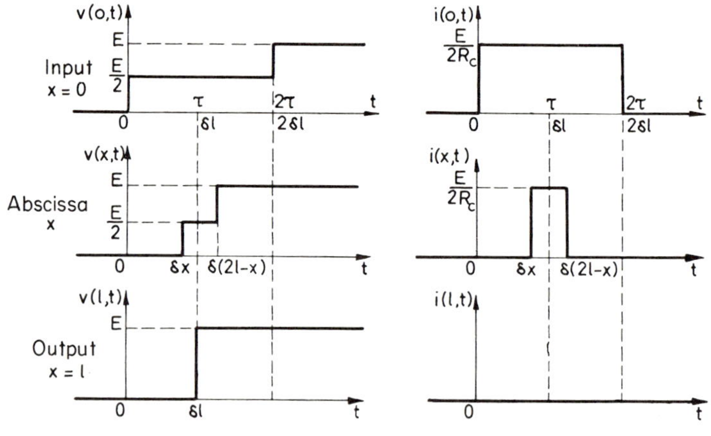

Fig. 190. Voltage and current waveforms for circuit of Fig. 189.

In this particular case, the voltage undergoes reflection without change of sign at the output, since the incident wave is E/2, as is the reflected wave, which allows the output voltage to jump from 0 to E at the instant δl at which a perturbation first reaches the output. On the other hand, the current is reflected with change of sign, so that the incident wave $E/2R_c$ is canceled by the reflected wave $-E/2R_c$, causing the output current to remain at zero.

Exercise 3

For the circuit of Fig. 191, we have $Z_0 = R_c$, hence $\Gamma_0 = 0$, $Z_l = 0$, hence $\Gamma_l = -1$, and $E(p) = E/p$. The basic equations are then

$$V(x, p) = \frac{E}{2p}\left(e^{-p\delta x} - e^{-p\delta(2l-x)}\right)$$

$$I(x, p) = \frac{E}{2pR_c}\left(e^{-p\delta x} + e^{-p\delta(2l-x)}\right)$$

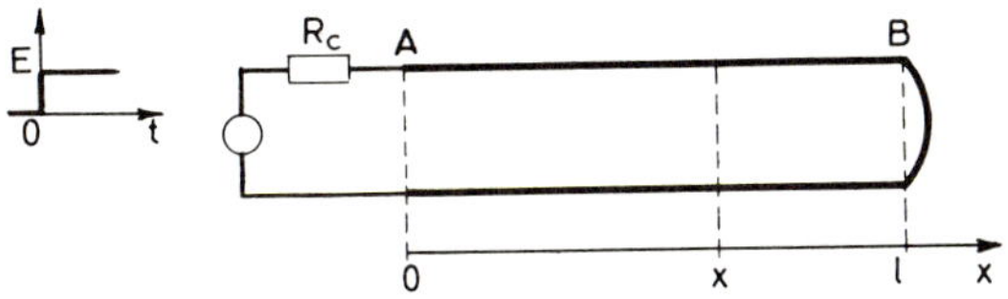

Fig. 191. Circuit for Exercise 3 of Chapter IV.

from which

$$v(x, t) = \frac{E}{2}\left[\Upsilon(t - \delta x) - \Upsilon(t - \delta(2l - x))\right]$$

$$i(x, t) = \frac{E}{2R_c}\left[\Upsilon(t - \delta x) + \Upsilon(t - \delta(2l - x))\right]$$

In particular, at the input, $x = 0$ and

$$v(0, t) = \frac{E}{2}\left[\Upsilon(t) - \Upsilon(t - 2\delta l)\right]$$

$$i(0, t) = \frac{E}{2R_c}\left[\Upsilon(t) + \Upsilon(t - 2\delta l)\right]$$

and at the output, $x = l$ and

$$v(l, t) = \frac{E}{2}\left[\Upsilon(t - \delta l) - \Upsilon(t - \delta l)\right] = 0$$

$$i(l, t) = \frac{E}{R_c}\Upsilon(t - \delta l)$$

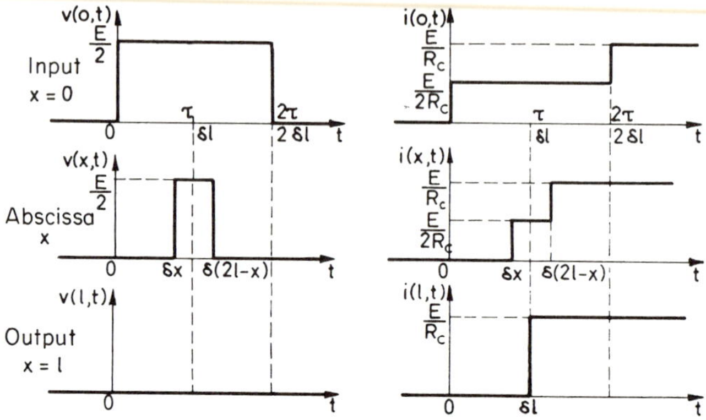

Fig. 192. Voltage and current waveforms in circuit of Fig. 191.

These are shown in Fig. 192.

In this case the voltage is reflected at the output with change of sign, so that the incident wave E/2 and the reflected wave −E/2 cancel, while the current is reflected without change of sign, allowing the incident wave $E/2R_c$ and reflected wave $E/2R_c$ to add, causing the output current to jump from 0 to E/R_c at time δl, as it must.

Exercise 4

(a) Replacing the generator in Exercise 1 by an arbitrary source $e(t)$ leads to the circuit of Fig. 193. We still have

$$\Gamma_0 = \frac{0 - R_c}{0 + R_c} = -1$$

since $Z_0 = 0$, and

$$\Gamma_l = \frac{R_c - R_c}{R_c + R_c} = 0$$

since $Z_l = R_c$. The basic equations (IV.3) and (IV.4) yield

$$V(x, p) = E(p) \frac{R_c}{Z_0 + R_c} e^{-p\delta x} = E(p) e^{-p\delta x}$$

$$I(x, p) = E(p) \frac{1}{Z_0 + R_c} e^{-p\delta x} = \frac{E(p)}{R_c} e^{-p\delta x}$$

from which

$$v(x, t) = e(0, t)\,\Upsilon(t - \delta x)$$

$$i(x, t) = \frac{e(0, t)}{R_c}\,\Upsilon(t - \delta x)$$

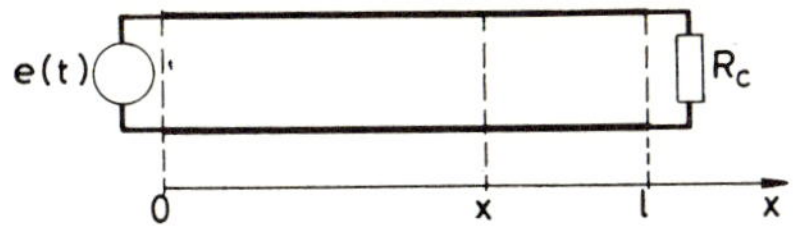

Fig. 193. Circuit for part (a) of Exercise 4 of Chapter IV.

Thus the voltage and current at time t, at an arbitrary point x along the line, are the same as the voltage and current at the input at time $t - \delta x$. This results from the fact that the load is matched to the line, and entirely absorbs all incident waves, so that there are no reflections.

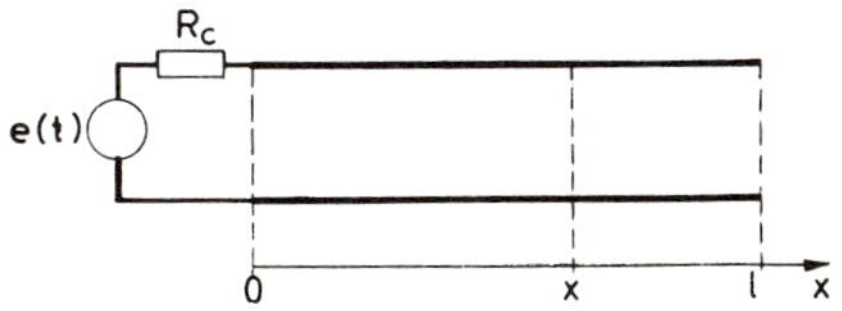

Fig. 194. Circuit for part (b) of Exercise 4 of Chapter IV.

(b) Replacing the generator in Exercise 2 by a source $e(t)$ leads to the circuit of Fig. 194. We have again

$$\Gamma_0 = \frac{R_c - R_c}{R_c + R_c} = 0$$

since $Z_0 = R_c$, and

$$\Gamma_l = \frac{Z_l - R_c}{Z_l + R_c} = 1$$

since $Z_l = \infty$. The basic equations yield

$$V(x, p) = \frac{E(p)}{2} [e^{-p\delta x} + e^{-p\delta(2l-x)}]$$

$$I(x, p) = \frac{E(p)}{2R_c} [e^{-p\delta x} - e^{-p\delta(2l-x)}]$$

from which

$$v(x, t) = \frac{1}{2} [e(t - \delta x)\Upsilon(t - \delta x) + e[t - \delta(2l - x)]\Upsilon[t - \delta(2l - x)]]$$

$$i(x, t) = \frac{1}{2R_c} [e(t - \delta x)\Upsilon(t - \delta x) - e[t - \delta(2l - x)]\Upsilon[t - \delta(2l - x)]]$$

The terms involving $\Upsilon(t - \delta x)$ correspond to the input state at $t = 0$, propagated without deformation to arrive at the point x at time $t = \delta x$.

The terms with $\Upsilon(t - \delta(2l - x))$ arise from propagation of the input state at $t = 0$ without deformation, to arrive at x at time $t = 2\delta l - \delta x$ after reflection at the end of the line. The signs of these terms (positive for voltage and negative for current), show that at the output the voltage reflection is without change of sign, while the current reflection is with change of sign.

Upon arriving again at the input, these waves are totally absorbed by the matched source load R_c. No energy is reflected back into the line at the source.

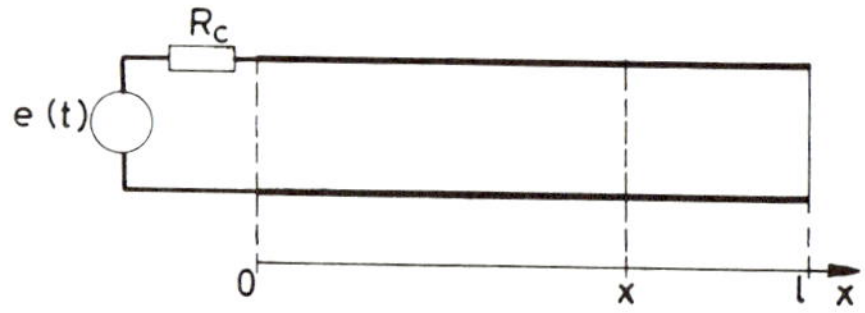

Fig. 195. Circuit for part (c) of Exercise 4 of Chapter IV.

(c) For Exercise 3 modified as in Fig. 195, we still have $Z_0 = R_c$, hence

$$\Gamma_0 = \frac{R_c - R_c}{R_c + R_c} = 0$$

and $Z_l = 0$, hence

$$\Gamma_l = \frac{0 - R_c}{0 + R_c} = -1$$

The basic equations become

$$V(x, p) = \frac{E(p)}{2}\,[e^{-p\delta x} - e^{-p\delta(2l-x)}]$$

$$I(x, p) = \frac{E(p)}{2R_c}\,[e^{-p\delta x} + e^{-p\delta(2l-x)}]$$

from which

$$v(x, t) = \frac{1}{2}\,[e(t - \delta x)\Upsilon(t - \delta x) - e[t - \delta(2l - x)]\Upsilon[t - \delta(2l - x)]]$$

$$i(x, t) = \frac{1}{2R_c}\,[e(t - \delta x)\Upsilon(t - \delta x) + e[t - \delta(2l - x)]\Upsilon[t - \delta(2l - x)]]$$

These equations have the same general interpretations as those for the lossless line, matched at the input, and open at the output, but in the present case, the voltage reflection at the output is with change of sign, and the current reflection is without change of sign.

Remark. It is also possible to reason as follows in these cases. A wave of any arbitrary shape can be approximated by a superposition of steps, as

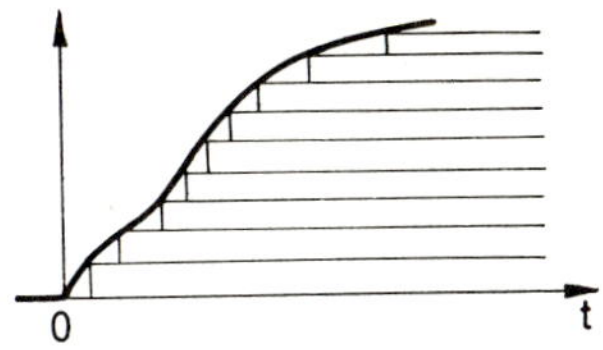

Fig. 196. Approximation of an arbitrary wave by a superposition of step functions.

indicated in Fig. 196. Thus, in the linear cases treated here, all results established for step waveforms remain valid for arbitrary waveforms.

Exercise 5

Consider the circuit of Fig. 197. Relation (IV.24), with

$$Z_0 = R_0; \qquad \Gamma_0 = \frac{R_0 - R_c}{R_0 + R_c} < 0; \qquad \Gamma_l = -1$$

eads to

$$i(l, t) = \frac{2}{R_0 + R_c} \sum_{n=0}^{\infty} \Gamma_0{}^n(-1)^n e[t - (2n + 1)\tau]\Upsilon[t - (2n + 1)\tau]$$

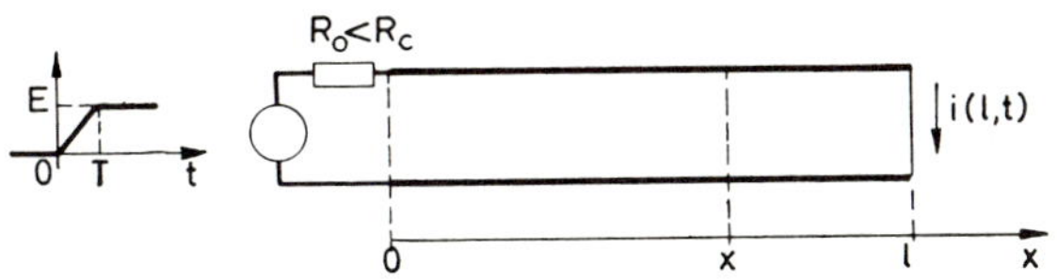

Fig. 197. Circuit for Exercise 5 of Chapter IV.

Here $\Gamma_0 < 0$ because $R_0 < R_c$. Setting

$$\Gamma = -\Gamma_0 = \frac{R_c - R_0}{R_c + R_0}$$

we have

$$\Gamma_0{}^n(-1)^n = \Gamma^n$$

which leads to

$$i(l, t) = \frac{2}{R_0 + R_c} \sum_{n=0}^{\infty} \Gamma^n e[t - (2n + 1)\tau]\Upsilon[t - (2n + 1)\tau]$$

This last equation shows that the current in the short circuit is the sum of a series of current ramps, with amplitudes decreasing as Γ^n and beginning at times $t = \tau, 3\tau, \ldots, (2n + 1)\tau, \ldots$. Figure 198 shows these ramps and their sum, in the case that $R_c = 50\,\Omega$, $R_0 = 10\,\Omega$, $T = 20$ nsec, $\tau = 4$ nsec,

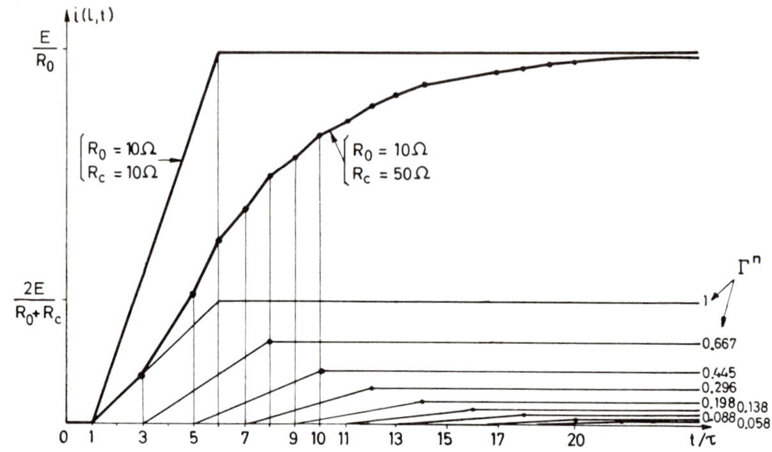

Fig. 198. Load current in circuit of Fig. 197 as a sum of ramp components.

from which

$$\Gamma = \frac{50 - 10}{50 + 10} = \frac{2}{3}$$

In the same figure is shown also the short-circuit current for the case that the generator is matched to the line, $R_0 = R_c$.

More generally, if the generator voltage $e(t)$ is some arbitrary waveform, rather than a ramp, the solution to this problem is again obtained by graphical summation of the terms of $i(l, t)$, which are again of the same waveform as $e(t)$. It is thus sufficient to have available a graph of $e(t)$. It is not particularly useful to have available $E(p)$.

Exercise 6

(1) Consider the circuit of Fig. 199. We have $E(p) = E/p$, $Z_0 = R_c$, $\Gamma_0 = 0$, $Z_l = R_1 + L_1 p$, and

$$\Gamma_l = \frac{(R_1 - R_c) + L_1 p}{(R_1 + R_c) + L_1 p}$$

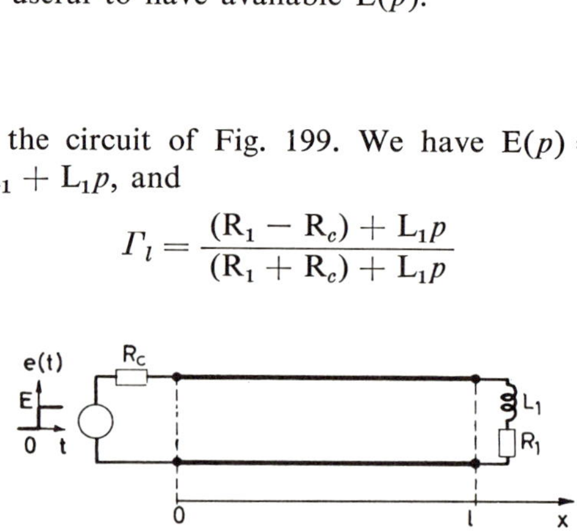

Fig. 199. Circuit for part (a) of Exercise 6 of Chapter IV.

In this case relations (IV.3) and (IV.4) yield

$$V(l,p) = \frac{E}{2p}\left[1 + \frac{(R_1 - R_c) + L_1 p}{(R_1 + R_c) + L_1 p}\right] e^{-\tau p} \qquad \text{with} \quad \tau = \delta l$$

$$I(l,p) = \frac{E}{2R_c p}\left[1 - \frac{(R_1 - R_c) + L_1 p}{(R_1 + R_c) + L_1 p}\right] e^{-\tau p}$$

$$V(l,p) = \frac{E}{2p}\left[1 + \frac{(R_1 + R_c) + L_1 p}{(R_1 + R_c) + L_1 p} - \frac{2R_c}{(R_1 + R_c) + L_1 p}\right] e^{-\tau p}$$

$$= \frac{E}{p}\left[1 - \frac{R_c}{(R_1 + R_c) + L_1 p}\right] e^{-\tau p}$$

Let us define $L_1/(R_1 + R_c) = T$. Then

$$V(l,\ p) = E\left[\frac{1}{p} - \frac{R_c/L_1}{p(p + 1/T)}\right] e^{-\tau p}$$

$$= E\left[\frac{1}{p} - \frac{R_c}{R_1 + R_c}\left(\frac{1}{p} - \frac{1}{p + 1/T}\right)\right] e^{-\tau p}$$

which has for inverse transform

$$v(l,\ t) = E\left[1 - \frac{R_c}{R_1 + R_c}\left(1 - e^{-(t-\tau)/T}\right)\right]\Upsilon(t - \tau)$$

In the same way the current can be found to be

$$i(l,\ t) = \frac{E}{R_1 + R_c}\left(1 - e^{-(t-\tau)/T}\right)\Upsilon(t - \tau)$$

Figure 200 shows the above current and voltage. Everything evolves as if the load were bridged directly across the generator, but delayed a time τ.

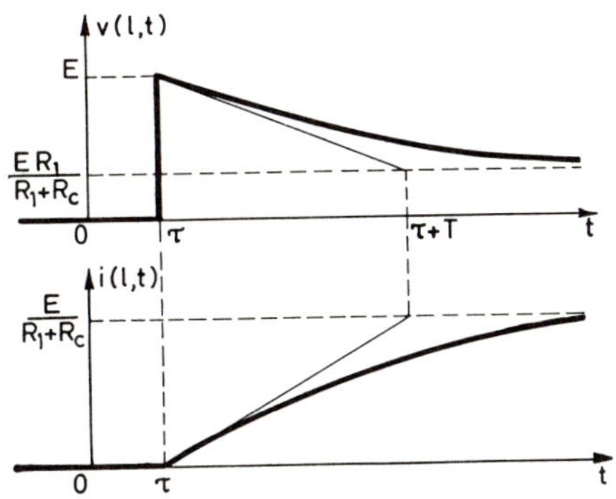

Fig. 200. Output voltage and current for circuit of Fig. 199.

(2) We have now the circuit of Fig. 201. This is a case of multiple reflections, with one of the reflection coefficients being a function of p:

$$\Gamma_l = \frac{Z_l - R_c}{Z_l + R_c} = \frac{L_1 p}{2R_c + L_1 p} = \frac{p}{p + 2R_c/L_1} = \frac{p}{p + 1/T}$$

where $T = L_1/2R_c$.

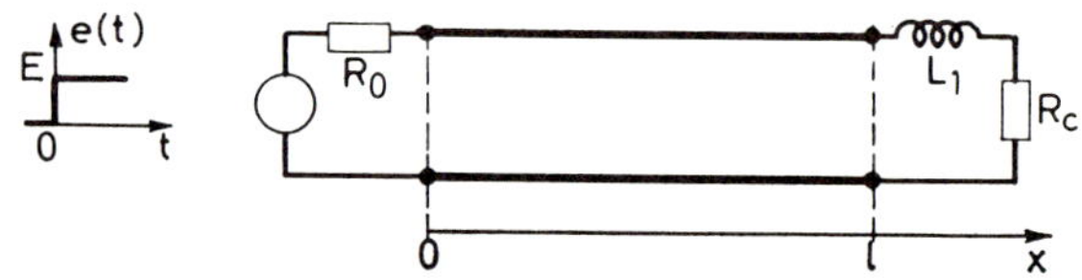

Fig. 201. Circuit for part (b) of Exercise 6 of Chapter IV.

Relation (IV.5) yields the output voltage as

$$\begin{aligned} V(l, p) &= \frac{E}{p} \frac{R_c}{R_0 + R_c} \left(1 + \frac{p}{p + 1/T}\right) \sum_{n=0}^{\infty} \Gamma_0{}^n \frac{p^n}{(p + 1/T)^n} e^{-(2n+1)\tau p} \\ &= E \frac{R_c}{R_0 + R_c} \left(\frac{1}{p} + \frac{1}{p + 1/T}\right) \sum_{n=0}^{\infty} \Gamma_0{}^n \frac{p^n}{(p + 1/T)^n} e^{-(2n+1)\tau p} \end{aligned}$$

To determine the inverse transform of this, it is necessary to invert terms of the form

$$F_n(p) = \left(\frac{1}{p} + \frac{1}{p + 1/T}\right) \frac{p^n}{(p + 1/T)^n} = \frac{p^{n-1}}{(p + 1/T)^n} + \frac{p^n}{(p + 1/T)^{n+1}}$$

which is to say

$$F_n(p) = F_{1n}(p) + F_{2n}(p)$$

where

$$F_{1n}(p) = \frac{p^{n-1}}{(p + 1/T)^n}, \quad F_{2n}(p) = \frac{p^n}{(p + 1/T)^{n+1}}$$

The function $F_{1n}(p)$ has a pole of order n at $p = -1/T$. The inverse $f_{1n}(t)$ of $F_{1n}(p)$ is the residue of $F_{1n}(p) \exp(pt)$ at $p = -1/T$,* which is to say

$$f_{1n}(t) = \frac{1}{(n-1)!} \left[\frac{d^{n-1}}{dp^{n-1}} (p^{n-1} e^{pt})\right]_{p=-1/T}$$

* See G. Metzger and J.-P. Vabre, "Electronique des Impulsions," Vol. I, "Circuits à Constantes Localisées." Masson, Paris, 1966.

Using Leibniz's formula, this becomes

$$f_{1n}(t) = \frac{1}{(n-1)!}[(n-1)(n-2)\ldots 2.1e^{pt}+(n-1)(n-1)(n-2)\ldots 2pte^{pt}$$

$$+\frac{(n-1)(n-2)}{2!}(n-1)(n-2)\ldots 3p^2t^2e^{pt}+\cdots+p^{n-1}t^{n-1}e^{pt}]_{p=-1/\mathrm{T}}$$

or finally

$$f_{1n}(t) = e^{-t/\mathrm{T}}\Big[1+(n-1)\Big(-\frac{t}{\mathrm{T}}\Big)+\frac{(n-1)(n-2)}{2!2!}\frac{t^2}{\mathrm{T}^2}$$

$$+\frac{(n-1)(n-2)(n-3)}{3!3!}\Big(-\frac{t}{\mathrm{T}}\Big)^3+\cdots+\frac{1}{(n-1)!}\Big(-\frac{t}{\mathrm{T}}\Big)^{n-1}\Big]$$

Similarly, $\mathrm{F}_{2n}(p)$ has a pole of order $n+1$ at $p=-1/\mathrm{T}$. Its inverse is thus

$$f_{2n}(t) = \frac{1}{n!}\Big[\frac{d^n}{dp^n}(p^n e^{pt})\Big]_{p=-1/\mathrm{T}}$$

$$f_{2n}(t) = \frac{1}{n!}[n(n-1)\ldots 2.1e^{pt}+n\cdot n(n-1)\ldots 2pte^{pt}$$

$$+\frac{n(n-1)}{2!}n(n-1)\ldots 3p^2t^2e^{pt}+\cdots+p^nt^ne^{pt}]_{p=-1/\mathrm{T}}$$

$$= e^{-t/\mathrm{T}}\Big[1+n\Big(-\frac{t}{\mathrm{T}}\Big)+\frac{n(n-1)}{2!2!}\frac{t^2}{\mathrm{T}^2}$$

$$+\frac{n(n-1)(n-2)}{3!3!}\Big(-\frac{t}{\mathrm{T}}\Big)^3+\cdots+\frac{1}{n!}\Big(-\frac{t}{\mathrm{T}}\Big)^n\Big]$$

Adding $f_{1n}(t)$ and $f_{2n}(t)$ we obtain

$$f_n(t) = e^{-t/\mathrm{T}}\Big[2-(2n-1)\frac{t}{\mathrm{T}}+\frac{(n-1)(2n-2)}{2!2!}\frac{t^2}{\mathrm{T}^2}-\frac{(n-1)(n-2)(2n-3)}{3!3!}\frac{t^3}{\mathrm{T}^3}$$

$$+\frac{(n-1)(n-2)(n-3)(2n-4)}{4!4!}\frac{t^4}{\mathrm{T}^4}+\cdots+\frac{(-1)^n}{n!}\Big(\frac{t}{\mathrm{T}}\Big)^n\Big]$$

From this it is then possible to calculate

$$v(l,t) = \mathrm{E}\frac{\mathrm{R}_c}{\mathrm{R}_0+\mathrm{R}_c}\sum_{n=0}^{\infty}\Gamma_0^n f_n[t-(2n+1)\tau]\Upsilon[t-(2n+1)\tau]$$

This exercise makes it plain that circuits in which multiple reflections are present, whether due to reactive elements or other causes, lead to complicated, but not impossible, calculations.

Remark. The presence of the factor $\Gamma_0{}^n$ in the above series causes the successive terms to decrease rapidly in magnitude. In practice, it suffices to calculate only the first few terms of the series. For $n = 4$ or 5, the duration of the transient regime is 8τ or 10τ.

Exercise 7

We consider here a line with no shunt losses, $G = 0$, and matched at the output, as in Fig. 202, so that $\Gamma_l = 0$. The internal impedance of the

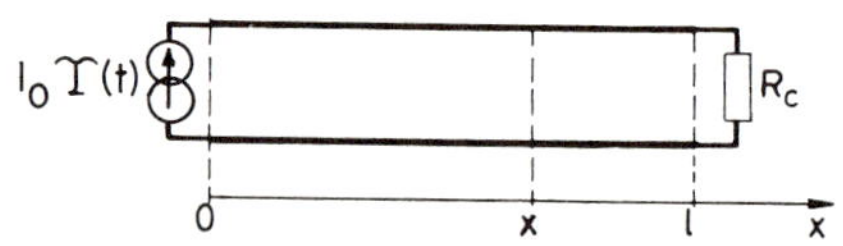

Fig. 202. Circuit for Exercise 7 of Chapter IV.

generator is assumed so large, $|Z_0| \gg |Z_c|$, that the source can be considered to be a current step, $E(p)/Z_0 = I_0/p$. In the beginning of the transient regime, we have Eq. (IV.28), which, using (IV.25) and (IV.27), becomes

$$\begin{aligned} V(x, p) &= \frac{I_0}{p} Z_c e^{-p\delta x} e^{-\alpha x} \\ &= \frac{I_0}{p} \left(R_c + \frac{1}{p} \frac{R}{2\delta} \right) e^{-p\delta x} e^{-Rx/2R_c} \\ &= I_0 e^{-Rx/2R_c} \left(\frac{R_c}{p} + \frac{1}{p^2} \frac{R}{2\delta} \right) e^{-p\delta x} \end{aligned}$$

The inverse transform is

$$v(x, t) = I_0 e^{-Rx/2R_c} [R_c \Upsilon(t - \delta x) + \frac{R}{2\delta} (t - \delta x) \Upsilon(t - \delta x)]$$

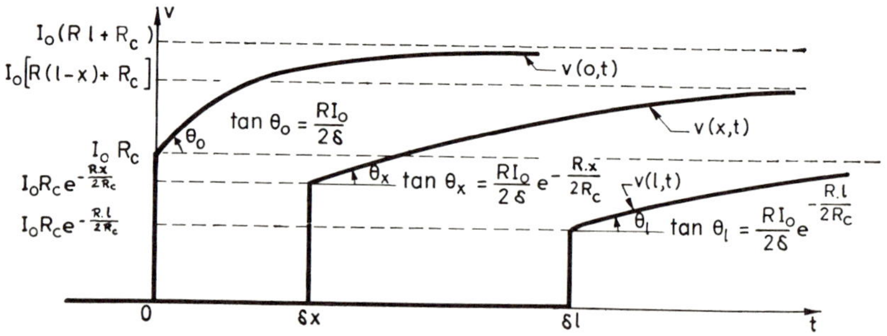

Fig. 203. Voltage waveforms in circuit of Fig. 202.

Thus at the beginning of the transient regime, the voltage increases as a ramp, the slope of which is a function of x and R_c. Note that from the slope of the ramp it is possible to determine the resistance per unit length R. The response curves of Fig. 203 are drawn using the above result for the initial part of the transient, and the steady-state solutions (asymptotes).

Exercise 8

(1) The characteristic resistance is $R_c = (L/C)^{1/2}$, just as for a lossless line. The propagation velocity is $u = 1/(LC)^{1/2} = c/\varepsilon_r^{1/2}$, from which $R_c = \varepsilon_r^{1/2}/cC$, where c is the velocity of light in vacuum, and C is the capacitance per unit length of line.

The line will be treated as a parallel-plate capacitor, neglecting edge effects. Thus

$$C = \varepsilon_0 \varepsilon_r \frac{a \times 1}{e}$$

so that

$$R_c = \frac{e}{\varepsilon_0 \varepsilon_r^{1/2} c a} = \frac{1}{\varepsilon_0 c}\left(\frac{e}{\varepsilon_r^{1/2} a}\right)$$

$$R_c = 120\pi \frac{e}{\varepsilon_r^{1/2} a}$$

With Teflon as dielectric,

$$\varepsilon_r = 2.25; \qquad \varepsilon_r^{1/2} = 1.5$$

$$R_c \approx 250\left(\frac{e}{a}\right)$$

So far as the attenuation is concerned, with Teflon dielectric the shunt losses are negligible, even at high frequency. Thus only series losses need be considered, and

$$\alpha = R/2R_c$$

After propagating along a length l of line, a signal will be attenuated by $\exp(-\alpha l) \approx 1 - \alpha l$, which is to say it will be decreased by a fraction αl. We have

$$\alpha l = \frac{Rl}{2R_c} = 2\varrho \frac{1}{ab} \frac{l}{2R_c} = \frac{\varrho l}{ab \times 250(e/a)} = \frac{\varrho l}{250eb}$$

The factor 2 in the third group is necessary since 2 unit lengths of material are involved per unit length of line, the forward and return paths. Thus finally

$$\alpha = \varrho/250eb$$

which is independent of the width a of the line.

(2) For the various cases, we have from above

$$R_c = 50\,\Omega; \qquad a/e = 5$$
$$R_c = 10\,\Omega; \qquad a/e = 25$$
$$R_c = 5\,\Omega; \qquad a/e = 50$$

The approximations made above are justified. Taking $e = 0.1$ mm, we have, respectively, $a = 0.5$, 2.5, and 5 mm, which are acceptable and realizable. Requiring $\alpha l \leq 5\%$ for $l = 0.20$ meter leads to

$$b \geq \left(\frac{\varrho l}{250e}\right)\frac{1}{0.05}$$

$$b \geq \frac{(1.7 \times 10^{-8}) \times 0.2}{250 \times (0.1 \times 10^{-3}) \times 0.05}$$

$$b \geq 2.75 \times 10^{-6} \quad \text{meters}$$

Thus we may take $b \approx 3\ \mu$.

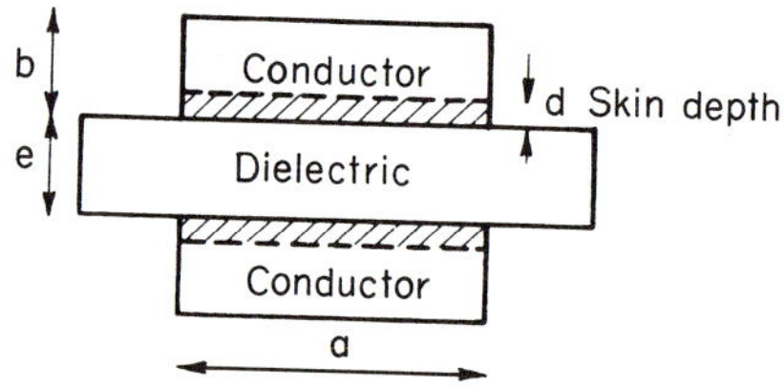

Fig. 204. Stripline considered in Exercise 8 of Chapter IV.

(3) Skin-effect losses arise in the propagation of a voltage step down the line of Fig. 204, giving rise to a time constant $T_{0.5}$. A time $110T_{0.5}$ is necessary for the step to reach 95% of its final value. We know that

$$T_{0.5} = K^2 l^2 / 4R_c^2$$

where $K = R(p)/p^{1/2}$. Let us use the method of the Appendix to Chapter IV. The length of the two conductors, each of length l, the potential difference $V(p)$, and the current are related by

$$V(p) = R(p)\, 2l\, I(p)$$

in transform notation. In the time domain,

$$v(t) = \varrho \frac{2l}{ad}\, i(t) = \varrho \frac{2l}{ad}\, ad\, j_0(t)$$

The depth d is defined such that the current density is constant and equal to $j_0(t)$, the current density at the surface. Thus

$$V(p) = \varrho 2l\, J_0(p)$$

But

$$J_0(p) = (\mu\sigma p)^{1/2} \frac{I(p)}{a} = \left(\frac{\mu}{\varrho}\, p\right)^{1/2} \frac{I(p)}{a}$$

Hence

$$V(p) = \frac{2l}{a}\, (\mu\varrho)^{1/2} p^{1/2} I(p)$$

and

$$K = 2(\varrho\mu)^{1/2}/a$$

Thus finally

$$T_{0.5} = \frac{\varrho\mu}{a^2}\, l^2\, \frac{1}{R_c^{\,2}}$$

Replacing R_c by $250(e/a)$ leads to the expression

$$T_{0.5} = \frac{\varrho\mu l^2}{(250)^2 e^2} = \frac{\varrho\mu_0\mu_r}{(250)^2}\, \frac{l^2}{e^2}$$

which is remarkable, in that it is independent of the width a and thickness b of the conductors. Thus $T_{0.5}$ depends only on the nature of the conductors, as expressed by ϱ and μ_r.

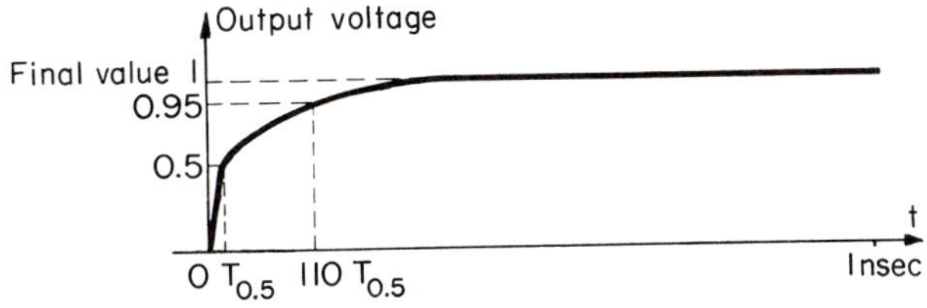

Fig. 205. Output voltage step response for a strip line, taking account of skin effect.

For copper, $\mu_r = 1$, $\varrho = 1.7 \times 10^{-8}\,\Omega$meter. For $l = 0.2$ meter and $e = 0.1 \times 10^{-3}$ meter, we have

$$T_{0.5} = \frac{1.7 \times 10^{-8} 4\pi 10^{-7} (0.2)^2}{(250)^2 (0.1 \times 10^{-3})^2} \approx 1.5 \times 10^{-12} \quad \text{sec}$$

or $T_{0.5} \approx 1.5$ psec. Thus $110 T_{0.5} < 10^{-9}$ sec (Fig. 205).

Exercise 9

(1) The line can be replaced by the equivalent self-inductance $L_{eq} = \tau R_c$, as in Fig. 206. The equivalent circuit in the transient regime is that of Fig. 206, where ΔV is the allowable variation of voltage and ΔI is the required current.

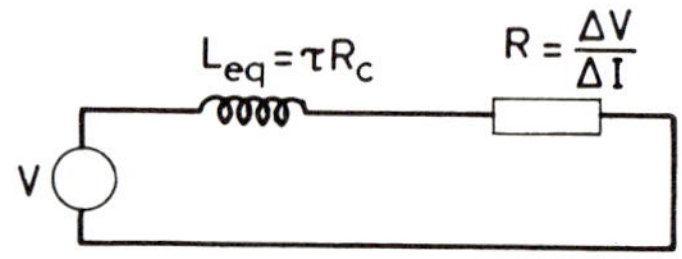

Fig. 206. Equivalent circuit for line in part (a) of Exercise 9 of Chapter IV.

(2) The equivalent circuit is modified by adding a capacitance C in parallel with the load, as in Fig. 207. The capacitor acts as a buffer, and initially supplies the current ΔI which the inductance L_{eq} does not allow to be established in the line sufficiently rapidly.

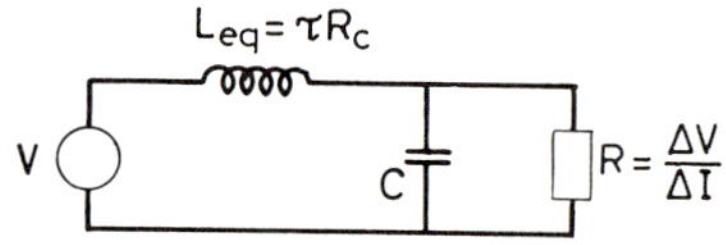

Fig. 207. Circuit for part (b) of Exercise 9 of Chapter IV.

(3) The circuit of Fig. 207 has a damping factor*

$$\zeta = \frac{1}{2R}\left(\frac{L}{C}\right)^{1/2}$$

For minimum response time without oscillations, the circuit should be critically damped, $\zeta = 1$. If some small oscillations are tolerable, the response time can be further improved by taking say $\zeta = 0.7$. The capacitance to be placed in parallel with the load can then be found as

$$C = \frac{L}{4R^2\zeta^2} = \frac{\tau R_c}{4R^2(0.7)^2} \approx \frac{\tau}{2}\,\frac{R_c}{R^2}$$

Remark. The usefulness of lines with low characteristic resistance R_c in the distribution of supply voltages through apparatus of large dimensions is clear. The smaller is R_c the smaller can be the capacitance C.

* See G. Metzger and J.-P. Vabre, "Electronique des Impulsions," Vol. I, "Circuits à Constantes Localisées, Chapter III. Masson, Paris, 1966.

For example, with a line having $R_c = 5\,\Omega$, and with $I = 1$ A, let us find the capacitance necessary to maintain ΔV less than 50 mV at the load, if the line has a length such that the propagation time is 5 nsec.

We have

$$R = \frac{\Delta V}{\Delta I} = \frac{50 \times 10^{-3}}{1} = 5 \times 10^{-2} \quad \Omega$$

The equivalent self-inductance of the line is

$$L_{eq} = \tau R_c = 5 \times 10^{-9} \times 5 = 25 \times 10^{-9} \quad \text{H}$$

Thus

$$C = \frac{L}{2R^2} = \frac{25 \times 10^{-9}}{2(5 \times 10^{-2})^2} = \frac{10^{-9}}{2 \times 10^{-4}} = \frac{10^{-5}}{2} \quad \text{F}$$

or $C = 5\,\mu$F. The same calculation made for a line with $R_c = 100\,\Omega$ rather than $5\,\Omega$ leads to $C = 100\,\mu$F.

Index

B

Bergeron, method of, 65–83
 with nonlinear elements, 73
 simultaneous observers, 74–83
 single observer, 68–73

C

Capacitance, equivalent, 137–138
Charging a line, *see* specific type of line
Circuit, equivalent lumped, 131–138
Coaxial line
 characteristic resistance of, 19
 inductance of, 17
 per-unit-length parameters of, 18–19, 102–103
 skin effect in, 130–131
Coefficient, reflection, 10–12, 23–28
 at discontinuity, 54–55
 current
  at input, 24
  at output, 24
 of current generator, 26
 of matched generator, 26
 of matched load, 25
 of open-circuited line, 26
 of short-circuited line, 26
 voltage
  at input, 10, 24
  at output, 11, 23
 of voltage generator, 28
Conductor
 forward, 1
 return, 1
Constant
 attenuation, 60
  of lossy line, 122
 propagation, 7
 time, skin-effect, 127–128
Constants
 of free space, 18
 per-unit-length, *see* Parameters, per-unit-length
Current
 conservation of, 2
 line, 3

D

Depth, skin
 operational, 140, 142
 real, 145–146
Diode, tunnel, 186
Discharging a line, *see* specific type of line
Discontinuous line, 81–83, 107–109
Dispersion, 158–160
Dispersionless line, 59–60
Distortionless line, 160

E

Effect
 proximity, 14
 skin, 12–13, 125–128, 138–146
Equations
 basic, for transmission line, 12, 58, 98
 general, for transmission line, 5–6
 of lossless line, 6–7
 telegraphers', 6

F

Function
  attenuation, 154
  phase, 154
  propagation, 7–9, 121–122, 151–152, 154
    of lossy line, 121–122

H

Heaviside condition, 124–125, 160
Homogeneity, 1

I

Impedance
  characteristic, 9–10, 99, 121, 152–153
  of line, 160–163
Inductance
  equivalent, 131–137
  of two-wire line, 16–17
Input, arbitrary, 92

L

Load
  capacitive, 90–91, 105–107
  distributed, 45–48
  inductive, 43–45, 86–88
  open circuit, *see* Open-circuited line
  short circuit, *see* Short-circuited line
Loss, dielectric, 15
Lossless line
  as transformer, *see* Transformer
  charging, 76–81
    with voltage step, 30, 113–116
  dispersion in, 158
  general equations of, 100–101
  impedance of, 163
  multiple reflections on, 109–119
  propagation constant of, 101–102
  traveling waves on, 21–25
Lossy line
  charging with voltage step, 31–33, 122–124
  dispersion in, 159–160
  Heaviside condition satisfied, 124–125
  in beginning of transient regime, 120–124
  traveling waves on, 57–61
Lumped elements in line, 84–91

M

Matching using stubs, 49–50, 174–175
Maxwell's equations, 139

N

Nonlinear problem, treated with Bergeron's method, 73

O

Open-circuited line
  charging with current ramp, 33–35
    with nonideal voltage step source, 35–37
    with voltage ramp, 116–119
    with voltage step, 29–33
  discharge with current ramp, 41–43
    with current step, 39–41

P

Parameters, per-unit-length, 4, 15–20, 151
  admittance, 151
  capacitance, 4
  conductance, 4, 17–18
  impedance, 151
  inductance, 4, 15–17
  resistance, 4, 15
Per-unit-length quantities, *see* Parameters, per-unit-length
Phase convention, 150
Propagation function, *see* Function, propagation

R

Radiation from line, 15
Reflection coefficient, *see* Coefficient, reflection
Reflections
  in lossless line, 109–119
  partial, 28
Resistance, characteristic, 7, 23, *see also* Impedance, characteristic

Resistance-loaded line
  charged
    by current ramp, 93
    by voltage ramp, 92–93
    by voltage step, 38–39, 69–73

S

Skin effect, *see* Effect, skin
State of line, 3
Strip line, 201–203
Stubs, matching, 49–50, 174–175
Symmetry of line, 3

T

Transform, Laplace, 7–12
Transformer, 163–166
  eighth-wave, 166
  half-wave, 164
  quarter-wave, 165
Transient regime, beginning of, 120
Transmission line, *see* specific type of line
Traveling waves
  on lossy line, 57–61
  method of, 21–25
  in sinusoidal steady state, 154–156
  tabular method, 51–54

V

Velocity, propagation, 7
Voltage, line, 3

W

Wave
  incident, 22
  reflected, 22
  traveling, *see* Traveling wave

ELECTRICAL SCIENCE

A Series of Monographs and Texts

Editors

Henry G. Booker
UNIVERSITY OF CALIFORNIA AT SAN DIEGO
LA JOLLA, CALIFORNIA

Nicholas DeClaris
UNIVERSITY OF MARYLAND
COLLEGE PARK, MARYLAND

Joseph E. Rowe. Nonlinear Electron-Wave Interaction Phenomena. 1965

Max J. O. Strutt. Semiconductor Devices: Volume I. Semiconductors and Semiconductor Diodes. 1966

Austin Blaquiere. Nonlinear System Analysis. 1966

Victor Rumsey. Frequency Independent Antennas. 1966

Charles K. Birdsall and William B. Bridges. Electron Dynamics of Diode Regions. 1966

A. D. Kuz'min and A. E. Salomonovich. Radioastronomical Methods of Antenna Measurements. 1966

Charles Cook and Marvin Bernfeld. Radar Signals: An Introduction to Theory and Application. 1967

J. W. Crispin, Jr., and K. M. Siegel (eds.). Methods of Radar Cross Section Analysis. 1968

Giuseppe Biorci (ed.). Network and Switching Theory. 1968

Ernest C. Okress (ed.). Microwave Power Engineering:
Volume 1. Generation, Transmission, Rectification. 1968
Volume 2. Applications. 1968

T. R. Bashkow (ed.). Engineering Applications of Digital Computers. 1968

Julius T. Tou (ed.). Applied Automata Theory. 1968

Robert Lyon-Caen. Diodes, Transistors, and Integrated Circuits for Switching Systems. 1969

M. Ronald Wohlers. Lumped and Distributed Passive Networks. 1969

Michel Cuenod and Allen E. Durling. A Discrete-Time Approach for System Analysis. 1969

K. Kurokawa. An Introduction to the Theory of Microwave Circuits. 1969

H. K. Messerle. Energy Conversion Statics. 1969

George Tyras. Radiation and Propagation of Electromagnetic Waves. 1969

Georges Metzger and Jean-Paul Vabre. Transmission Lines with Pulse Excitation. 1969

C. L. Sheng. Threshold Logic. 1969

Dale M. Grimes. Electromagnetism and Quantum Theory. 1969

Robert O. Harger. Synthetic Aperture Radar Systems: Theory and Design. 1969

In Preparation

M. A. Lampert and P. Mark. Current Injection in Solids.

W. V. T. Rusch and P. D. Potter. Analysis of Reflector Antennas.

TRANSMISSION LINES WITH PULSE EXCITATION

This treatise features a time-dependent treatment of transmission lines emphasizing numerical and graphical solutions.

The authors begin with a consideration of the unit step function and of superpositions of unit step functions to describe an arbitrary pulse with no dispersion and little or no loss. Using the Laplace transform, they proceed to the propagation of an arbitrary pulse on an arbitrary line with dispersion and loss. They conclude with the generalization of the considered techniques to the behavior of a line carrying a sinusoidal signal.

Topics covered include:

- General Equations for Transmission Lines
- The Method of Traveling Waves
- The Bergeron Diagram
- Line Transients Using Operational Calculus
- Lines in the Sinusoidal Steady State

This work will be of interest to electrical engineers, engineering students, and physicists.